Basic Principles of Wastewater Treatment

Biological Wastewater Treatment Series

The *Biological Wastewater Treatment* series is based on the book *Biological Wastewater Treatment in Warm Climate Regions* and on a highly acclaimed set of best selling textbooks. This international version is comprised by six textbooks giving a state-of-the-art presentation of the science and technology of biological wastewater treatment.

Titles in the *Biological Wastewater Treatment* series are:

Volume 1: *Wastewater Characteristics, Treatment and Disposal*
Volume 2: *Basic Principles of Wastewater Treatment*
Volume 3: *Waste Stabilisation Ponds*
Volume 4: *Anaerobic Reactors*
Volume 5: *Activated Sludge and Aerobic Biofilm Reactors*
Volume 6: *Sludge Treatment and Disposal*

Biological Wastewater Treatment Series

VOLUME TWO

Basic Principles of Wastewater Treatment

Marcos von Sperling

Department of Sanitary and Environmental Engineering
Federal University of Minas Gerais, Brazil

Publishing
LONDON • NEW YORK

Published by IWA Publishing, Alliance House, 12 Caxton Street, London SW1H 0QS, UK

Telephone: +44 (0) 20 7654 5500; Fax: +44 (0) 20 7654 5555; Email: publications@iwap.co.uk
Website: **www.iwapublishing.com**

First published 2007
© 2007 IWA Publishing

Copy-edited and typeset by Aptara Inc., New Delhi, India

Disclaimer
The information provided and the opinions given in this publication are not necessarily those of IWA or of the editors, and should not be acted upon without independent consideration and professional advice. IWA and the editors will not accept responsibility for any loss or damage suffered by any person acting or refraining from acting upon any material contained in this publication.

British Library Cataloguing in Publication Data
A CIP catalogue record for this book is available from the British Library

Library of Congress Cataloguing-in-Publication Data
A catalogue record for this book is available from the Library of Congress

ISBN: 1 84339 162 7
ISBN 13: 9781843391623

Contents

Preface

The present series of books has been produced based on the book "*Biological wastewater treatment in warm climate regions*", written by the same authors and also published by IWA Publishing. The main idea behind this series is the subdivision of the original book into smaller books, which could be more easily purchased and used.

The implementation of wastewater treatment plants has been so far a challenge for most countries. Economical resources, political will, institutional strength and cultural background are important elements defining the trajectory of pollution control in many countries. Technological aspects are sometimes mentioned as being one of the reasons hindering further developments. However, as shown in this series of books, the vast array of available processes for the treatment of wastewater should be seen as an incentive, allowing the selection of the most appropriate solution in technical and economical terms for each community or catchment area. For almost all combinations of requirements in terms of effluent quality, land availability, construction and running costs, mechanisation level and operational simplicity there will be one or more suitable treatment processes.

Biological wastewater treatment is very much influenced by climate. Temperature plays a decisive role in some treatment processes, especially the natural-based and non-mechanised ones. Warm temperatures decrease land requirements, enhance conversion processes, increase removal efficiencies and make the utilisation of some treatment processes feasible. Some treatment processes, such as anaerobic reactors, may be utilised for diluted wastewater, such as domestic sewage, only in warm climate areas. Other processes, such as stabilisation ponds, may be applied in lower temperature regions, but occupying much larger areas and being subjected to a decrease in performance during winter. Other processes, such as activated sludge and aerobic biofilm reactors, are less dependent on temperature, as a result of the higher technological input and mechanisation level. The main purpose of this series of books is to present the technologies for urban wastewater treatment as applied to the specific condition of warm temperature, with the related implications in terms of design and operation. There is no strict definition for the range of temperatures that fall into this category, since the books always present how to correct parameters, rates and coefficients for different temperatures. In this sense, subtropical and even temperate climate are also indirectly covered, although most of the focus lies on the tropical climate.

Another important point is that most warm climate regions are situated in developing countries. Therefore, the books cast a special view on the reality of these countries, in which simple, economical and sustainable solutions are strongly demanded. All technologies presented in the books may be applied in developing countries, but of course they imply different requirements in terms of energy, equipment and operational skills. Whenever possible, simple solutions, approaches and technologies are presented and recommended.

Considering the difficulty in covering all different alternatives for wastewater collection, the books concentrate on off-site solutions, implying collection and transportation of the wastewater to treatment plants. No off-site solutions, such as latrines and septic tanks are analysed. Also, stronger focus is given to separate sewerage systems, although the basic concepts are still applicable to combined and mixed systems, especially under dry weather conditions. Furthermore, emphasis is given to urban wastewater, that is, mainly domestic sewage plus some additional small contribution from non-domestic sources, such as industries. Hence, the books are not directed specifically to industrial wastewater treatment, given the specificities of this type of effluent. Another specific view of the books is that they detail biological treatment processes. No physical-chemical wastewater treatment processes are covered, although some physical operations, such as sedimentation and aeration, are dealt with since they are an integral part of some biological treatment processes.

The books' proposal is to present in a balanced way theory and practice of wastewater treatment, so that a conscious selection, design and operation of the wastewater treatment process may be practised. Theory is considered essential for the understanding of the working principles of wastewater treatment. Practice is associated to the direct application of the concepts for conception, design and operation. In order to ensure the practical and didactic view of the series, 371 illustrations, 322 summary tables and 117 examples are included. All major wastewater treatment processes are covered by full and interlinked design examples which are built up throughout the series and the books, from the determination of the wastewater characteristics, the impact of the discharge into rivers and lakes, the design of several wastewater treatment processes and the design of the sludge treatment and disposal units.

The series is comprised by the following books, namely: *(1) Wastewater characteristics, treatment and disposal; (2) Basic principles of wastewater treatment; (3) Waste stabilisation ponds; (4) Anaerobic reactors; (5) Activated sludge and aerobic biofilm reactors; (6) Sludge treatment and disposal.*

Volume 1 (*Wastewater characteristics, treatment and disposal*) presents an integrated view of water quality and wastewater treatment, analysing wastewater characteristics (flow and major constituents), the impact of the discharge into receiving water bodies and a general overview of wastewater treatment and sludge treatment and disposal. Volume 1 is more introductory, and may be used as teaching material for undergraduate courses in Civil Engineering, Environmental Engineering, Environmental Sciences and related courses.

Volume 2 (*Basic principles of wastewater treatment*) is also introductory, but at a higher level of detailing. The core of this book is the unit operations and processes associated with biological wastewater treatment. The major topics covered are: microbiology and ecology of wastewater treatment; reaction kinetics and reactor hydraulics; conversion of organic and inorganic matter; sedimentation; aeration. Volume 2 may be used as part of postgraduate courses in Civil Engineering, Environmental Engineering, Environmental Sciences and related courses, either as part of disciplines on wastewater treatment or unit operations and processes.

Volumes 3 to 5 are the central part of the series, being structured according to the major wastewater treatment processes (*waste stabilisation ponds, anaerobic reactors, activated sludge and aerobic biofilm reactors*). In each volume, all major process technologies and variants are fully covered, including main concepts, working principles, expected removal efficiencies, design criteria, design examples, construction aspects and operational guidelines. Similarly to Volume 2, volumes 3 to 5 can be used in postgraduate courses in Civil Engineering, Environmental Engineering, Environmental Sciences and related courses.

Volume 6 (*Sludge treatment and disposal*) covers in detail sludge characteristics, production, treatment (thickening, dewatering, stabilisation, pathogens removal) and disposal (land application for agricultural purposes, sanitary landfills, landfarming and other methods). Environmental and public health issues are fully described. Possible academic uses for this part are same as those from volumes 3 to 5.

Besides being used as textbooks at academic institutions, it is believed that the series may be an important reference for practising professionals, such as engineers, biologists, chemists and environmental scientists, acting in consulting companies, water authorities and environmental agencies.

The present series is based on a consolidated, integrated and updated version of a series of six books written by the authors in Brazil, covering the topics presented in the current book, with the same concern for didactic approach and balance between theory and practice. The large success of the Brazilian books, used at most graduate and post-graduate courses at Brazilian universities, besides consulting companies and water and environmental agencies, was the driving force for the preparation of this international version.

In this version, the books aim at presenting consolidated technology based on worldwide experience available at the international literature. However, it should be recognised that a significant input comes from the Brazilian experience, considering the background and working practice of all authors. Brazil is a large country with many geographical, climatic, economical, social and cultural contrasts, reflecting well the reality encountered in many countries in the world. Besides, it should be mentioned that Brazil is currently one of the leading countries in the world on the application of anaerobic technology to domestic sewage treatment, and in the post-treatment of anaerobic effluents. Regarding this point, the authors would like to show their recognition for the Brazilian Research Programme on Basic Sanitation (PROSAB), which, through several years of intensive, applied, cooperative research has led to the consolidation of anaerobic treatment and aerobic/anaerobic post-treatment, which are currently widely applied in full-scale plants in Brazil. Consolidated results achieved by PROSAB are included in various parts of the book, representing invaluable and updated information applicable to warm climate regions.

Volumes 1 to 5 were written by the two main authors. Volume 6 counted with the invaluable participation of Cleverson Vitorio Andreoli and Fernando Fernandes, who acted as editors, and of several specialists, who acted as chapter authors: Aderlene Inês de Lara, Deize Dias Lopes, Dione Mari Morita, Eduardo Sabino Pegorini, Hilton Felício dos Santos, Marcelo Antonio Teixeira Pinto, Maurício Luduvice, Ricardo Franci Gonçalves, Sandra Márcia Cesário Pereira da Silva, Vanete Thomaz Soccol.

Many colleagues, students and professionals contributed with useful suggestions, reviews and incentives for the Brazilian books that were the seed for this international version. It would be impossible to list all of them here, but our heartfelt appreciation is acknowledged.

The authors would like to express their recognition for the support provided by the Department of Sanitary and Environmental Engineering at the Federal University of Minas Gerais, Brazil, at which the two authors work. The department provided institutional and financial support for this international version, which is in line with the university's view of expanding and disseminating knowledge to society.

Finally, the authors would like to show their appreciation to IWA Publishing, for their incentive and patience in following the development of this series throughout the years of hard work.

Marcos von Sperling
Carlos Augusto de Lemos Chernicharo

December 2006

The author

Marcos von Sperling
PhD in Environmental Engineering (Imperial College, Univ. London, UK).
Associate professor at the Department of Sanitary and Environmental Engineering, Federal University of Minas Gerais, Brazil. Consultant to governmental and private companies in the field of water pollution control and wastewater treatment.
marcos@desa.ufmg.br

1

Microbiology and ecology of wastewater treatment

1.1 INTRODUCTION

Biological wastewater treatment, as the name suggests, occurs entirely by biological mechanisms. These biological processes reproduce, in a certain way, the natural processes that take place in a water body after a wastewater discharge. In a water body, organic matter is converted into inert mineralised products by purely natural mechanisms, characterising the *self-purification* phenomenon. In a wastewater treatment plant the same basic phenomena occur, but the difference is that there is the introduction of technology. This technology has the objective of making the *purification* process develop under **controlled conditions** (operational control) and at **higher rates** (more compact solution).

The understanding of the microbiology of sewage treatment is therefore essential for the optimisation of the design and operation of biological treatment systems. In the past, engineers designed the treatment works based essentially on empirical criteria. In the last few decades, the multidisciplinary character of Sanitary and Environmental Engineering has been recognised, and the biologists have brought fundamental contributions for the understanding of the process. The rational knowledge has expanded, together with the decrease in the level of empiricism, allowing the systems to be designed and operated with a more solid base. The result has brought an increase in the efficiency and a reduction in the costs.

The main organisms involved in sewage treatment are bacteria, protozoa, fungi, algae and worms. Their characterisation is presented in Section 1.2. This chapter also covers sewage treatment from a biological and ecological (study of the communities involved) points of view. Recognising the great importance of the bacteria in the conversion process of the organic matter, a more detailed description of them is given in the present chapter.

1.2 MICROORGANISMS PRESENT IN WATER AND WASTEWATER

Microbiology is the branch of biology that deals with microorganisms. In terms of water quality, the microorganisms play an essential role, due to their large predominance in certain environments, their action in wastewater purification processes and their association with water borne diseases. Microorganisms can only be observed microscopically.

Some microorganism groups have properties in common with plants whilst others have some animal characteristics. In the past, the classification of living creatures used to be according to the two main kingdoms, **Plants** and **Animals**, and the microorganisms were present in each of these two large subdivisions.

Subsequently, however, biologists have adopted a more practical division, placing microorganisms in the separate kingdoms of the **Monera** (simpler creatures, without a separate nucleus, such as bacteria, cyanobacteria and archaea) and the **Protists** (simple creatures, but with a separate nucleus, such as algae, fungi and protozoa). There are still other possible subdivisions into other kingdoms, but these are not important for the objectives of this book.

The basic difference between the monera/protists and the other organisms (plants and animals) is the high level of cellular differentiation found in the plants and animals. This means that, in monera and protist organisms, the cells of a single individual are morphologically and functionally similar, which reduces its adaptation and development capacity. However, in organisms with cellular differentiation, a functional division occurs. In the higher organisms, the differentiated cells (generally of the same type) combine into larger or smaller groups, called tissue. The tissues constitute the organs (e.g. lungs), and these form the systems (e.g. respiratory system). The level of cellular differentiation is therefore an indication of the developmental level of a species.

Table 1.1 presents the basic characteristics of the kingdoms in the living world, while Table 1.2 lists the main characteristics of the various groups that comprise the monera and protist kingdoms.

Protists have the nucleus of the cell confined by a nuclear membrane (algae, protozoa and fungi), being characterised **eukaryotes**. Monera have the nucleus disseminated in the protoplasm (bacteria, cyanobacteria and archaea), being characterised as **prokaryotes**. In general, the eukaryotes present a higher level of internal differentiation and may be unicellular or multicellular. The viruses were not included in the above classification because of their totally specific characteristics. Cyanobacteria were previously called blue-green algae. Archaea are similar to bacteria in size and basic cell components. However, their cell wall, cell material

Table 1.1. Basic characteristics of the kingdoms of the living world

Characteristics	Monera and Protists	Plants	Animals
Cell	Unicellular or multicellular	Multicellular	Multicellular
Cellular differentiation	Non-existent	High	High
Energy source	Light, organic matter or inorganic matter	Light	Organic matter
Chlorophyll	Absent or present	Present	Absent
Movement	Immobile or mobile	Immobile	Mobile
Cell wall	Absent or present	Present	Absent

Table 1.2. Basic characteristics of the main groups of microorganisms
(monera and protists)

	Monera (prokaryotes)		Protists (eukaryotes)		
Characteristic	Bacteria	Cyanobacteria	Algae	Protozoa	Fungi
Nuclear membrane	Absent	Absent	Present	Present	Present
Photosynthesis	Minority	Majority	Yes	No	No
Movement	Some	Some	Some	Mobile	Immobile

Source: adapted from La Riviére (1980)

and RNA composition are different. Archaea are important in anaerobic processes (mainly methanogenesis).

1.3 BIOLOGICAL CELLS

Generally, the majority of living cells are very similar. A short description of their main components is presented below (La Riviére, 1980; Tchobanoglous & Schroeder, 1985).

The cells generally have as an external boundary a *cell membrane*. This membrane is flexible and functions as a selective barrier between what is contained inside the cell and the external environment. The membrane is semi-permeable and therefore exerts an important role in selecting the substances that can leave or enter the cell. However, bacteria, algae, fungi, and plants have yet another external layer called *cell wall*. This is generally composed of a rigid material that gives structural form to the cell, even offering protection against mechanical impacts and osmotic alterations. It is believed that this wall is not semi-permeable and therefore does not exert a role in the regulation of the consumption of dissolved substances in the surrounding medium. In some bacteria the cell wall can even be involved by another external layer, generally of a gelatinous material, called *capsule* (with defined limits) or *gelatinous layer* (when diffused). In the case that the individual cells present motility, they usually have *flagella* or *cilia*.

The interior of the cell contains *organelles* and a colloidal suspension of proteins, carbohydrates and other complex forms of organic matter, constituting the *cytoplasm*.

Each cell contains *nucleic acids*, a genetic material vital for reproduction. Ribonucleic acid (RNA) is important for the synthesis of proteins and is found in

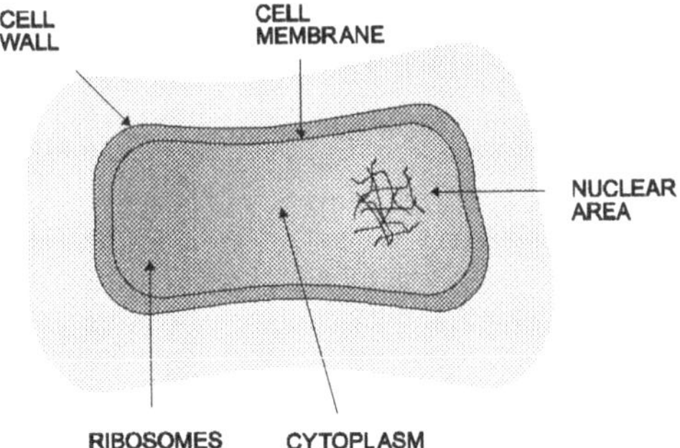

Figure 1.1.   Simplified schematic representation of a bacterial cell

the ribosome present in the cytoplasm. The prokaryotic cells, such as those from bacteria, contain only a *nuclear area*, while the eukaryotic cells have a *nucleus* enclosed within a membrane. The nucleus (or nuclear area) is rich in deoxyribonucleic acids (DNA), which contain the genetic information necessary for the reproduction of all the cell components. The cytoplasm of the prokaryotic cells frequently contains DNA in small structures called plasmids.

A simplified schematic representation of a bacterial cell is presented in Figure 1.1.

1.4 ENERGY AND CARBON SOURCES FOR MICROBIAL CELLS

All living creatures need (a) energy (b) carbon and (c) nutrients (nitrogen, phosphorus, sulphur, potassium, calcium, magnesium, etc.) for functions of growth, locomotion, reproduction and others.

In terms of the *carbon source*, there are two fundamental organism types:

> • **Autotrophic organisms**. Carbon source: *carbon dioxide* (CO_2);
> • **Heterotrophic organisms**. Carbon source: *organic matter.*

In terms of the *energy source*, there are two basic organism types:

> • **Phototrophic organisms**. Energy source: *light*;
> • **Chemotrophic organisms**. Energy source: *energy from chemical reactions.*

The combinations between these four types are shown in Table 1.3.

In most of the sewage treatment processes (with the exception of facultative and maturation ponds), light does not penetrate significantly in the liquid contained in

Table 1.3. General classification of microorganisms based on sources of energy and carbon

Classification	Energy source	Carbon source	Representative organisms
Photoautotrophs	Light	CO_2	Higher plants, algae, photosynthetic bacteria
Photoheterotrophs	Light	Organic matter	Photosynthetic bacteria
Chemoautotrophs	Inorganic matter	CO_2	Bacteria
Chemoheterotrophs	Organic matter	Organic matter	Bacteria, fungi, protozoa, animals

Source: Tchobanoglous and Schroeder (1985); Metcalf & Eddy (1991)

the biological reactors, due to the high turbidity of the liquid. Because of this, the presence of microorganisms that have light as energy sources (photoautotrophs and photoheterotrophs) is extremely limited. Therefore, the organisms of real importance in this case are the *chemoautotrophs* (responsible, for example, for nitrification) and the *chemoheterotrophs* (responsible for most of the reactions that occur in biological treatment). For simplicity, the latter ones will be subsequently named only *heterotrophs*.

1.5 METABOLISM OF MICROORGANISMS

The chemical processes that simultaneously take place in the cell are jointly called *metabolism*, and can be divided into two categories (La Riviére, 1980):

* *Dissimilation or catabolism*: reactions of energy production, in which substrate decomposition occurs;
* *Assimilation* or *anabolism*: reactions that lead to the formation of cellular material (growth), using the energy released in the dissimilation.

In a simplified way, the organisms grow and reproduce themselves at the expense of the energy released in the dissimilation. In dissimilation, the energy stored in chemical form in the organic compounds (substrate) is released and converted in the assimilation in cellular material. The net growth is the result of the balance between the anabolism (positive) and the catabolism (negative).

In both categories, the chemical transformations occur in a sequence of diverse and intricate intermediate reactions, each catalysed by a specific type of enzyme. Most of the enzymes are located inside the cell: these are called intracellular enzymes or *endoenzymes*. However, some enzymes are released in the external medium and are designated as extracellular enzymes or *exoenzymes*. Their importance is associated with the fact that they lead to hydrolysis reactions outside the cell, in the liquid medium, converting large and complex substrate molecules into smaller and simpler molecules, which can then pass through the cell membrane to become available for consumption by the cell.

Table 1.4. Main characteristics of the oxidative and fermentative catabolism

Characteristic	Oxidative catabolism (respiration)	Fermentative catabolism (fermentation)
Electron donor	Organic matter	Oxidised organic matter
Electron acceptor	External: inorganic compound (oxygen, nitrate or sulphate)	Internal: reduced organic matter
Number of final products resulting from the organic matter	One (CO_2)	At least two (CO_2 and CH_4)
Form of carbon in the final product	Oxidised inorganic carbon (CO_2)	Oxidised inorganic carbon (CO_2) + Reduced organic carbon (CH_4)
Oxidation state of carbon in the final product	$4+$ (CO_2)	$4+$ (CO_2) and $4-$ (CH_4)

The removal of the organic matter from sewage occurs through the process of dissimilation or catabolism. The two types of catabolism of interest in sewage treatment are *oxidative catabolism* (oxidation of the organic matter) and *fermentative catabolism* (fermentation of the organic matter) (van Haandel and Lettinga, 1994):

- *Oxidative catabolism*: redox reaction in which an oxidising agent present in the medium (oxygen, nitrate or sulphate) oxidises the organic matter.
- *Fermentative catabolism*: there is no oxidant. The process occurs due to the rearrangement of the electrons in the fermented molecule in such a way that at least two products are formed. Generally, there is the need of various fermentation sequences for the products to be stabilised, that is, be no longer susceptible to fermentation.

The main characteristics of the oxidative and fermentative catabolism are presented in Table 1.4. The concept of electron acceptor is explained in Section 1.6.

1.6 ENERGY GENERATION IN MICROBIAL CELLS

As seen in Section 1.5, the generation of energy in the microbial cells can be accomplished, depending on the microorganism, by means of respiration (oxidative catabolism) or fermentation (fermentative catabolism).

The name respiration is not restricted to the processes that involve oxygen consumption. In general, oxidation implies the loss of one or more electrons from the oxidised substance (in oxidation, the substance gives in negative charges in the form of electrons when passing to a higher oxidation state). The oxidised substance can be the organic matter, as well as reduced inorganic compounds – both are therefore **electron donors**. The electrons taken from the oxidised molecule are transferred through complicated biochemical reactions with the help of enzymes to another inorganic compound (oxidising agent), which receives the generic denomination

Table 1.5. Typical electron acceptors in the oxidation reactions in sewage treatment (listed in decreasing order of energy release)

Conditions	Electron acceptor	Form of the acceptor after the reaction	Process
Aerobic	*Oxygen* (O_2)	H_2O	Aerobic metabolism
Anoxic	*Nitrate* (NO_3^-)	Nitrogen gas (N_2)	Nitrate reduction (denitrification)
Anaerobic	*Sulphate* (SO_4^{2-})	Sulphide (H_2S)	Sulphate reduction
	Carbon dioxide (CO_2)	Methane (CH_4)	Methanogenesis

of **electron acceptor**. As a result, the electron acceptor has its oxidation state reduced.

The main electron acceptors used in respiration are listed in Table 1.5 in a decreasing order of energy release.

When various electron acceptors are available in the medium, the microorganisms use the one that produces the highest quantity of energy. For this reason, the *dissolved oxygen* is used first and, after it is exhausted, the system stops being **aerobic**. If there is *nitrate* available in the liquid medium (which is not always the case), the organisms that are capable of using nitrate in their respiration start to do it, converting the nitrate to nitrogen gas (denitrification). These conditions receive an specific name, being designated as **anoxic** (absence of dissolved oxygen but presence of nitrates). When the nitrates are finished, strict **anaerobic** conditions occur. In these, *sulphates* are used and reduced to sulphides, and *carbon dioxide* is converted into methane. While there are substances with greater energy release, the other ones are not used (Arceivala, 1981).

The methanogenesis stage can occur in two pathways. The first is the oxidative process of the *hydrogenotrophic methanogenesis* (production of methane from hydrogen), in which carbon dioxide acts as an electron acceptor and is reduced to methane. This pathway is less important in terms of the global conversion, but can be made by practically all methanogenic organisms. The second pathway is *acetotrophic methanogenesis* (production of methane from acetate), in which the organic carbon, in the form of acetate (acetic acid) is converted into methane. This pathway is responsible for most of the conversions, although it is accomplished by few bacteria species (Lubberding, 1995).

There are organisms that are functionally adapted to the various respiration conditions. The main ones are:

- **Strict aerobic organisms**: use only *free oxygen* in their respiration
- **Facultative organisms**: use *free oxygen* (preferably) or *nitrate* as electron acceptors
- **Strict anaerobic organisms**: use *sulphate* or *carbon dioxide* as electron acceptors and cannot obtain energy through aerobic respiration

Owing to the release of more energy through the aerobic reactions than through the anaerobic reactions, the aerobic organisms reproduce themselves and stabilise the organic matter faster than the anaerobes. Because of the reproduction rate of the aerobic organisms being greater, the sludge generation is also greater.

The main reactions for the generation of energy that occur in aerobic, anoxic and anaerobic conditions are:

- Aerobic conditions:

$$\boxed{C_6H_{12}O_6 + 6\,O_2 \text{ -------------> } 6\,CO_2 + 6\,H_2O} \qquad (1.1)$$

- Anoxic conditions: *nitrate reduction (denitrification)*

$$\boxed{2\,NO_3^- + 2\,H^+ \text{ ------------> } N_2 + 2{,}5\,O_2 + H_2O} \qquad (1.2)$$

- Anaerobic conditions: *sulphate reduction*

$$\boxed{CH_3COOH + SO_4^{2-} + 2\,H^+ \text{ ------------> } H_2S + 2\,H_2O + 2\,CO_2} \qquad (1.3)$$

- Anaerobic conditions: *CO_2 reduction (hydrogenotrophic methanogenesis)*

$$\boxed{4\,H_2 + CO_2 \text{ ------------> } CH_4 + 2\,H_2O} \qquad (1.4)$$

- Anaerobic conditions: *acetotrophic methanogenesis*

$$\boxed{CH_3COOH \text{ ------------> } CH_4 + CO_2} \qquad (1.5)$$

Figure 1.2 illustrates the main routes of organic matter decomposition in the presence of different electron acceptors.

The sequence of transformations that occurs in sewage treatment is a function of the electron acceptor and the oxidation state of the compounds, measured by its oxidation–reduction potential (expressed in millivolts). Figure 1.3 illustrates these reactions.

The oxidation state of the compound determines the maximum quantity of energy available through it. The more reduced the compound, the more energy it contains. The objective of the energetic metabolism is to conserve as much energy as possible in a form available for a cell. The maximum energy available from the oxidation of a substrate is the difference between its energetic content (given by the oxidation state) and the energetic content of the final products of the reaction (also given by their oxidation state at the end of the reaction) (Grady and Lim, 1980).

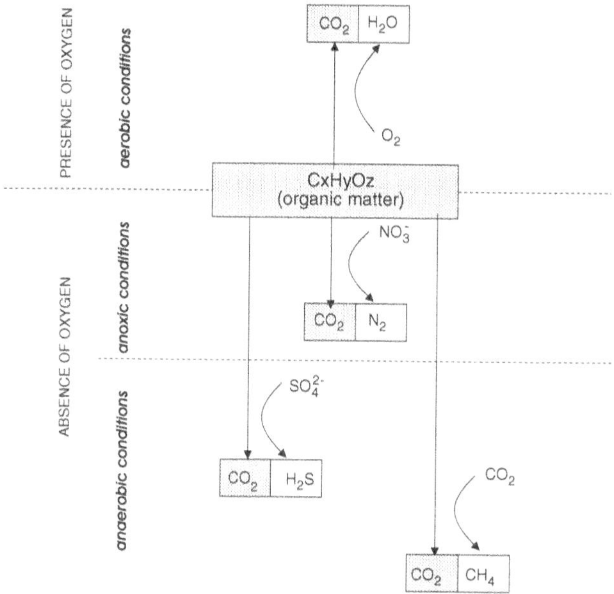

Figure 1.2.   Main routes of organic matter decomposition in the presence of different electron acceptors (modified from Lubberding, 1995)

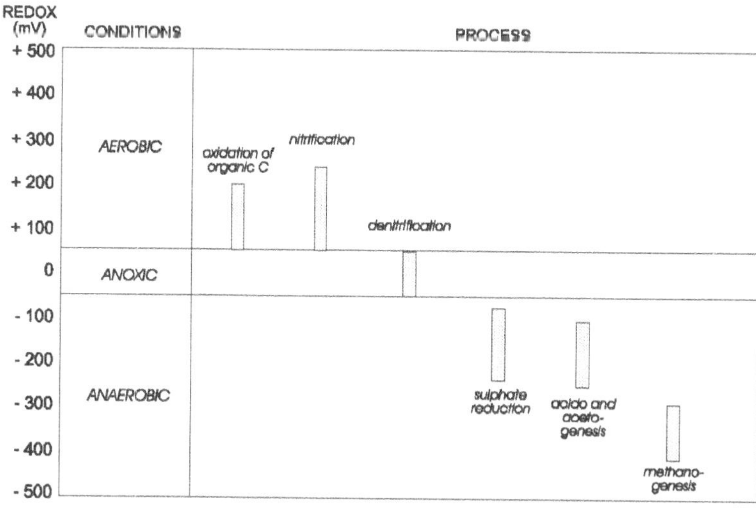

Figure 1.3.   Transformation sequence in sewage treatment, as a function of the electron acceptor and the redox potential (adapted from Eckenfelder and Grau, 1992)

The following points apply:

- The greater the oxidation state of the final product, the greater the energy release. The carbon in CO_2 is at its higher state of oxidation. Therefore oxidation reactions that oxidise the carbon in the substrate completely to CO_2 (aerobic respiration) release more energy than the reactions that produce, for example, ethanol (fermentation).
- The lower the oxidation state of the substrate, the greater the energy release. For instance, the oxidation of acetic acid to CO_2 releases less energy than the oxidation of ethanol to CO_2, because the carbon in the acetic acid is at a higher oxidation state than in ethanol.
- CO_2 can never act as an energy source, because its carbon is at the highest possible state of oxidation (CO_2 cannot be oxidised).

1.7 ECOLOGY OF BIOLOGICAL WASTEWATER TREATMENT

1.7.1 Introduction

The role played by microorganisms in sewage treatment depends on the process being used. In facultative ponds, algae have a fundamental function related to the production of oxygen by photosynthesis. The design of ponds is done in such a way as to optimise the presence of algae in the liquid medium and to obtain an adequate balance between bacteria and algae. In anaerobic treatment systems, the conditions are favourable or even exclusive for the development of microorganisms functionally adapted to the absence of oxygen. In this case, the acidogenic and methanogenic organisms are essential.

The microbial mass involved in the aerobic processes consists mainly of bacteria and protozoa. Other organisms, such as fungi and rotifers, can also be found, but their importance is lower. The capacity of fungi to survive in reduced pH ranges and with little nitrogen makes them important in the treatment of certain industrial wastewaters. However, fungi with a filamentous structure can deteriorate the sludge settleability, thus reducing the efficiency of the process. Rotifers are efficient in the consumption of dispersed bacteria and small particles of organic matter. Their presence in the effluent indicates an efficient biological purification process (Metcalf & Eddy, 1991). Generally, it can be said that the species diversity of the various microorganisms in the biomass is low.

Figure 1.4 presents a sequence of the relative predominance of the main microorganisms involved in aerobic sewage treatment. The ecological interactions in the microbial community cause the increase in the population of a group of microorganisms to be followed by the decline of another group, in view of the selective characteristics of the medium in transformation. Immediately after the introduction of sewage into the biological reactor, the remaining BOD (organic matter) is at its maximum level. The number of bacteria is still reduced, and protozoa of the amoeba type can be found. These are inefficient in the competition

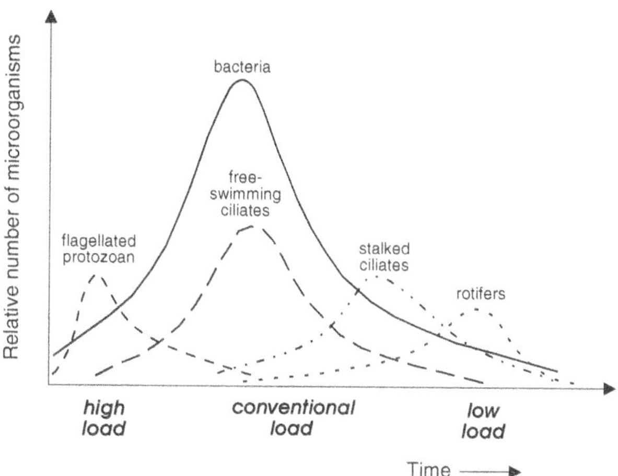

Figure 1.4.   Sequence of the relative predominance of the microorganisms in sewage treatment (adapted from König, 1990; Metcalf & Eddy, 1991)

for the available food, being found mainly at the start-up of reactors. Due to the great availability of the substrate, the bacterial population grows. The amoebas are then substituted by flagellated protozoa that, due to their motility, are more efficient in the competition for the available food. These flagellated protozoa are characteristic of *high load* systems. With the passing of time and the decrease of the available organic matter, ciliate protozoa substitute the flagellated ones, since the former are capable of surviving with lower food concentrations. This point characterises the operation of *conventional load* systems, where a large number of free-living ciliates are present together with the maximum number of bacteria and a low concentration of organic matter (remaining BOD). In long retention periods, which are characteristic of *low load* systems, the available organic matter is at a minimum and the bacteria are consumed by ciliates and rotifers (König, 1990).

The sections below describe the two main groups involved in the conversion of organic matter: bacteria and protozoa, with greater emphasis being given to the first group.

1.7.2 Bacteria

Bacteria are unicellular prokaryotic (absence of a defined nucleus) microorganisms that live isolated or in colonies. The classification of bacteria according to the shape includes the categories listed in Table 1.6.

Table 1.6. Categories of bacteria according to shape

Name	Shape	Size
Cocci (singular, coccus)	spheroid	0.5 to 3.0 μm in diameter
Bacilli (singular, bacillus)	rod	0.3 to 1.5 μm in width (or diameter) 1.0 to 10.0 μm in length
Spirilla (singular, spirillum)	spiral	<50 μm in length
Vibrios	curved rod	0.6 to 1.0 μm in width (or diameter) 2.0 to 6.0 μm in length
Various	filamentous	>100 μm in length

The bacteria have a more or less rigid cell wall and may or may not present flagella for locomotion. Their reproduction is principally by binary fission, besides the formation of spores and sexual reproduction (minority) (Branco, 1976; Metcalf & Eddy, 1991).

Bacteria constitute the largest and most important group in biological wastewater treatment systems. Considering that the main function of a treatment system is the removal of BOD, the heterotrophic bacteria are the main agents of this mechanism. In addition of playing the role of conversion of the organic matter, the bacteria have the property to agglomerate themselves in structural units such as flocs, biofilms or granules, which have important implications in wastewater treatment (see Section 1.7.4).

Besides the removal of the carbonaceous organic matter, sewage treatment can also incorporate other objectives, which depend on specific groups of bacteria. Thus, the following phenomena can take place:

- *Conversion of ammonia to nitrite (nitrification)*: chemoautotrophic bacteria
- *Conversion of nitrite to nitrate (nitrification)*: chemoautotrophic bacteria
- *Conversion of nitrate to nitrogen gas (denitrification)*: facultative chemo-heterotrophic bacteria

The cellular structure of bacteria was presented in Section 1.3 and illustrated in Figure 1.1. Approximately 80% of the bacterial cell is composed of water and 20% of dry matter. Of this dry matter, around 90% is organic and 10% inorganic. Widely used formulas for the characterisation of the approximate cell composition are (Metcalf & Eddy, 1991):

$$C_5H_7O_2N \qquad \text{(without phosphorus in the formula)}$$

$$C_{60}H_{87}O_{23}N_{12}P \qquad \text{(with phosphorus in the formula)}$$

In any of the two formulations, the C:H:O:N ratio is the same. An important aspect is that all of these components should be obtained from the medium, and the absence of any of them could limit the growth of the bacterial population.

The utilisation by the bacteria of the substrate available in the medium depends on the relative size of the particle. The two main fractions of the organic matter in the wastewater are (a) *easily biodegradable fraction* and (b) *slowly biodegradable fraction*. In a typical domestic sewage, most of the organic matter in soluble form

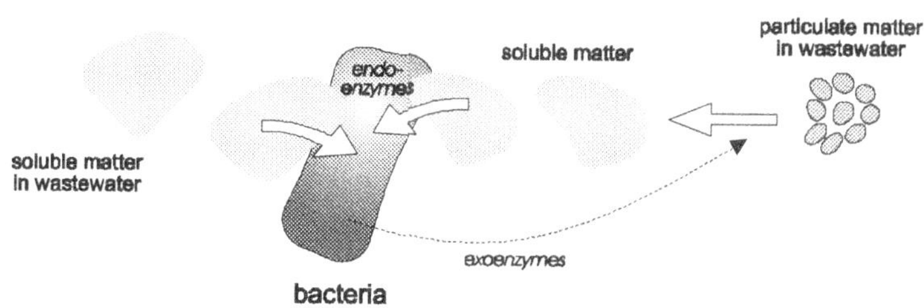

Figure 1.5.   Mechanisms of assimilation of soluble matter and conversion of particulate matter into soluble matter

is easily degradable. Owing to their small dimensions, soluble compounds can penetrate the bacterial cell through their cellular membranes. Inside the cell, the soluble organic matter is consumed with the aid of *endoenzymes*. Organic compounds of larger dimensions and more complex formulas (particulate or suspended organic matter), should undergo a transformation process outside the cells, leading to smaller molecules, which can be assimilated by the bacteria. This action is accomplished with the aid of *exoenzymes* in a reaction of *hydrolysis*. In the hydrolysis, energy consumption is not taken into account, and there is no use of electron acceptors. The final product of hydrolysis is present in an easily biodegradable form, penetrating through the cellular membrane inside the cell, where it is consumed in a similar way as the soluble matter (IAWPRC, 1987) (see Figure 1.5).

The environmental requirements for the bacteria vary with the species. For example, bacteria involved in the nitrification process (chemoautotrophic bacteria) are much more sensitive to environmental conditions than the usual heterotrophic bacteria involved in the stabilisation of the carbonaceous organic matter. In general, the optimal growth rate for bacterial growth occurs within relatively limited ranges of temperature and pH, even though their survival can occur within much wider ranges.

Temperatures below the optimal level have a greater influence in the growth rate compared to temperatures above the optimal level. Depending on temperature range, bacteria can be classified as *psycrophilic, mesophilic* or *thermophilic*. The typical temperature ranges for each of these categories are presented in Table 1.7.

pH is also an important factor in bacterial growth. Most bacteria do not support pH values above 9.5 or below 4.0, and the optimal value is around neutrality (6.5 to 1.5) (Metcalf & Eddy, 1991).

1.7.3 Protozoa

Most of the protozoa group comprises unicellular eukaryotic microorganisms without a cell wall. Although they have no cellular differentiation, some have a relatively complex structure with some differentiated regions within the cell for the

Table 1.4. Temperature ranges for optimal
bacterial growth

Type	Temperature (°C)	
	Range	Ideal
Psycrophilic	−10 to 30	12 to 18
Mesophilic	20 to 50	25 to 40
Thermophilic	35 to 75	55 to 65

Source: Metcalf & Eddy, 1991

undertaking of different functions. The majority are represented by strictly aerobic or facultative heterotrophic organisms. Their reproduction occurs by binary fission. Protozoa are usually larger than bacteria and can feed on them. This makes the protozoa group an important level in the food web, allowing that larger organisms feed indirectly on the bacteria, which would otherwise be an inaccessible form of food. Depending on some structural characteristics and on the mode of motility, the protozoa can be divided into various groups. Those of principal interest are the following: amoebas, flagellates and free-swimming and stalked ciliates (Branco, 1976; La Riviére, 1980). Some species are pathogenic.

In terms of the role of protozoa in biological wastewater treatment, the following are of importance:

- Consumption of organic matter;
- Consumption of free bacteria;
- Participation in floc formation.

The last aspect, related to the contribution to the formation of flocs, seems to be a mechanism of lower importance (La Riviére, 1977). The first two aspects (consumption of organic matter and of free bacteria) depend on the feeding mode of the protozoa, which varies with its type, as seen below (Horan, 1990):

- *Flagellates.* Use of soluble organic matter by diffusion or active transport. In this feeding mode, bacteria are more efficient in the competition.
- *Amoebas and ciliates.* Formation of a vacuole around the solid particle (that can include bacteria), through a process called phagocytosis. The organic fraction of the particle is then utilised after an enzymatic action inside the vacuole (inside the cell).
- *Ciliates* (principally). Predation of bacteria, algae and other ciliated and flagellated protozoa.

Although the protozoa contribute to the removal of the organic matter in sewage, their main role in treatment (by processes such as activated sludge) is by the predatory activity that they exert on bacteria freely suspended in the liquid medium (La Riviére, 1977). Hence, bacteria that are not part of the floc, but are dispersed in the medium are not normally removed in the final sedimentation. As a result, they contribute to the deterioration of the final effluent in terms of suspended solids, organic matter (from the bacteria themselves) and even pathogens. Therefore, the

TYPES OF BIOMASS GROWTH

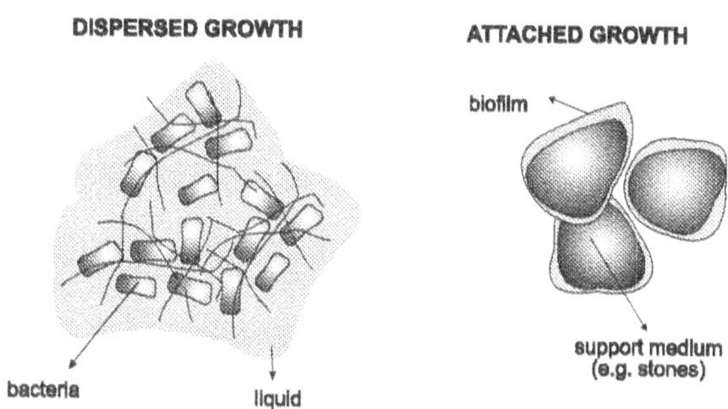

Figure 1.6. Typical examples of biomass growth

action of protozoa on bacteria contributes to improving the quality of the final effluent (Horan, 1990).

The free-swimming ciliates have greater food requirements than the stalked ciliates, because a large part of their energy is spent in locomotion. The predominance of the stalked ciliates occurs after the decline of the free-swimming ciliates population, when they can feed on the bacteria available in the floc.

1.7.4 Suspended and attached biomass growth

With relation to the structural formation of the biomass, biological sewage treatment processes can be divided into the basic configurations listed below (see Figure 1.6). The list is organised according to the prevailing mechanism, although mechanisms of attached and dispersed growth can occur simultaneously.

- **Dispersed growth**: the biomass grows in a dispersed form in the liquid medium, without any supporting structure
  Systems:
  - stabilisation ponds and variants
  - activated sludge and variants
  - upflow anaerobic sludge blanket reactors (receiving wastewaters containing suspended solids)
- **Attached growth**: the biomass grows attached to a support medium, forming a biofilm. The support medium can be immersed in the liquid medium or receive continuous or intermittent liquid discharges. The support medium can be a solid natural (stones, sand, soil) or artificial (plastic) material or consist of an agglomerate of the biomass itself (granules).
  Systems with a solid support for attachment:
  - trickling filters
  - rotating biological contactors

- submerged aerated biofilters
- anaerobic filters
- land disposal systems

Systems with the support for attachment consisting of the agglomerated biomass.

- upflow anaerobic sludge blanket reactors (receiving predominantly soluble sewage)

Even though the principles of biological treatment are the same for both biomass support systems, the treatment kinetics are influenced by the intervenience of specific aspects. The greater theoretical development in terms of modelling is with respect to aerobic treatment with dispersed growth. This results from the facts that there have been for many years a larger number of researches directed to the activated sludge process and that the formulation of dispersed-growth models is, in a way, simpler than for attached-growth systems.

1.7.5 The biological floc in dispersed-growth systems

In some treatment processes, such as activated sludge, the organisms concentrate and form a broader structural unit that is called a *floc*. Although microorganisms are the agents in BOD removal, the floc in activated sludge plays an essential role in the organic matter removal process. It is not only the property of the heterotrophic organisms in stabilising organic matter that makes the activated sludge process efficient. Also of fundamental importance is the property shown by the main microorganisms to organise themselves in the structural unit of a floc, which is capable of being separated from the liquid by the simple physical mechanism of sedimentation, in separate settling units. This separation allows the final effluent to be clarified (with reduced concentrations of suspended organic matter). The quality of the final effluent is therefore characterised by low values of soluble BOD (removed in the reactor) and of suspended BOD (flocs removed in the final settling unit). The flocculation mechanism, which in water treatment is reached at the expense of adding chemical products, occurs by entirely natural mechanisms in biological sewage treatment.

The floc represents a heterogeneous structure that contains the adsorbed organic matter, inert material from the sewage, microbial material produced, and alive and dead cells. The size of the floc is regulated by the balance between the forces of cohesion and shear stress caused by the artificial aeration and agitation (La Riviére, 1977). Among the microorganisms that constitute the floc, besides the bacteria and protozoa, fungi, rotifers, nematodes and occasionally even insect larvae can be found (Branco, 1978).

The floc matrix can rapidly absorb up to 40% of the soluble and particulate BOD entering the biological reactor through ionic interactions. The particulate material is hydrolysed by exoenzymes before it is absorbed and metabolised by the bacteria. Considering that the size of a floc varies between 50 and 500 μm, there will be a marked gradient of BOD and oxygen concentrations from the external

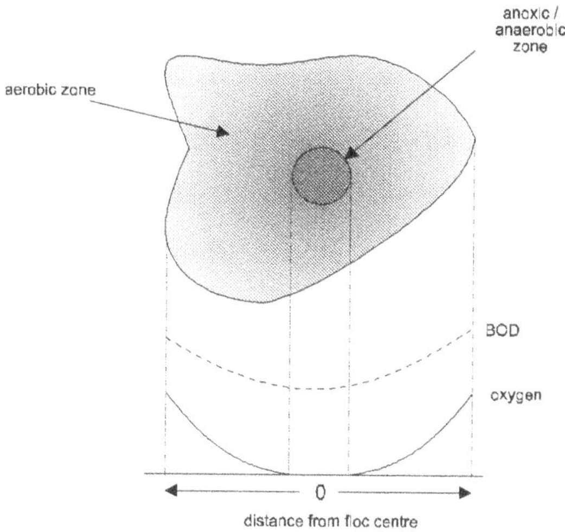

Figure 1.7.   BOD and oxygen gradients along a typical floc (adapted from Horan, 1990).

border of the floc (larger values) to the centre (where very low BOD values and zero DO values can be found). Consequently, in the direction of the centre of the floc, the bacteria become deprived of nutrient sources, what reduces their viability (Horan, 1990). When analysing the availability of oxygen or nutrients in the liquid medium, their possible absence inside the floc must be taken into consideration. This supports the fact that, for instance, in many cases anoxic conditions may be assumed, even though a small concentration of DO (0.5 mg/L) in the liquid medium may still be found. Although the liquid medium is not deprived of oxygen, a large part of the floc is. In this situation, the anoxic conditions are prevalent in the interior of the floc, and the biochemical reactions take place as if in the absence of oxygen. Figure 1.7 illustrates the BOD and oxygen gradients along the floc.

The conditions that lead to the microbial growth in the form of flocs instead of cells freely suspended in the liquid medium are not fully known. A plausible hypothesis for the structure of the floc is that the filamentous bacteria exert the function of a structural matrix, to which the floc-forming bacteria adhere. It is believed that the attachment occurs through exopolysaccharides, present in the form of a capsule or gelatinous layer. In the past this phenomena was attributed only to *Zoogloea ramigera,* but there are indications that the production of the gelatinous layer occurs through various genera, including *Pseudomonas.* The continuous production of these exopolymers results in the adherence of other microorganisms and colloidal particles and, as a consequence, the floc size increases. Finally, the protozoa adhere and colonise the floc, and there is some evidence that they also

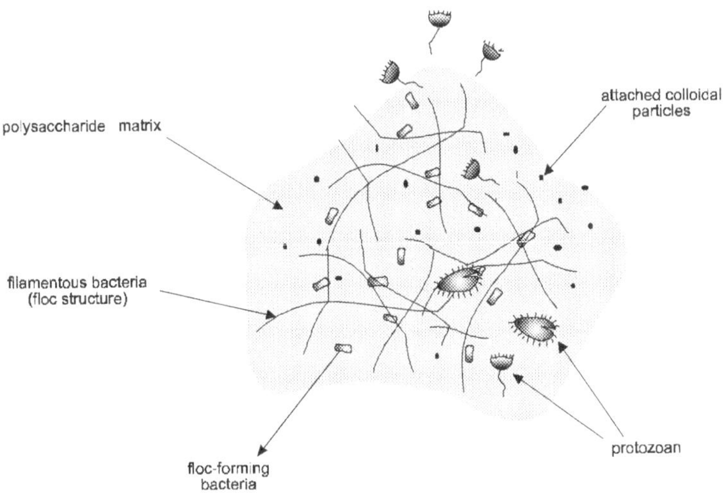

Figure 1.8.   Typical structure of an activated sludge floc (adapted from Horan, 1990).

excrete a viscous mucus that contributes to the cohesion of the floc (Horan, 1990). Figure 1.8 shows an schematics of a typical structure of an activated sludge floc.

The balance between the filamentous and floc-forming organisms is delicate, and a good part of the operational success of an activated sludge plant depends on it (see also Chapter 39). Three basic conditions can occur (Horan, 1990):

- *Equilibrium between filamentous and floc-forming organisms.* Good settleability and thickening properties of the sludge.
- *Predominance of floc-forming organisms.* There is insufficient rigidity in the floc, generating a small and weak floc, with poor settleability. This results in the so-called pin-point floc.
- *Predominance of filamentous organisms.* The filaments extend themselves outside the floc, impeding the adherence of other flocs. Therefore, after sedimentation, the flocs occupy an excessively large volume, which can bring problems in the operation of the secondary sedimentation tank, causing a deterioration in the quality of the final effluent. Such a condition is called sludge bulking.

1.7.6 Biofilm in attached growth systems

Immobilisation is the attachment of the microorganisms to a solid or suspended supporting medium. The immobilisation has the advantage of enabling a high biomass concentration to be retained in the reactor for long time periods. Although practically all the microorganisms have the potential to adhere to a supporting

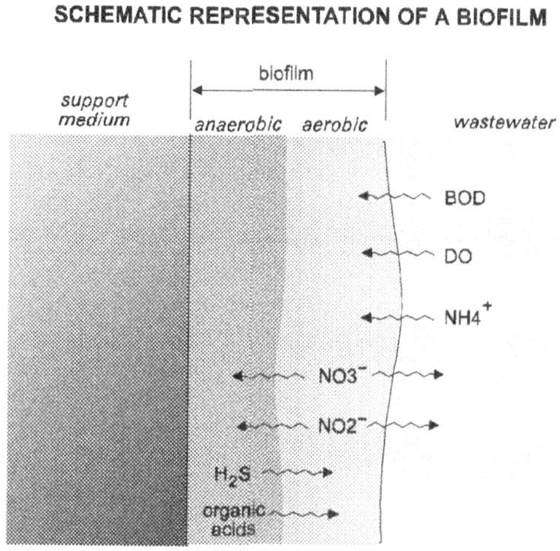

Figure 1.9.   Schematic representation of a biofilm (adapted from Iwai and Kitao, 1994)

medium through the production of extracellular polymers that allow physical–chemical attachment, it is only recently that the technological application of cellular sorption processes is being employed in a wider and optimised scale in various biotechnological processes and in sewage treatment (Lubberding, 1995).

The attachment is influenced by cell-to-cell interactions, by the presence of polymer molecules on the surface and by the composition of the medium (Rouxhet and Mozes, 1990).

In the biofilm, the compounds necessary for bacterial development, such as organic matter, oxygen and micronutrients, are adsorbed onto the surface. After adhering, they are transported through the biofilm through diffusion mechanisms, where they are metabolised by the microorganisms. Colloidal or suspended solids cannot diffuse through the biofilm and need to be hydrolysed to smaller molecules. The final metabolic products are transported in the opposite direction, to the liquid phase (Iwai and Kitao, 1994). Figure 1.9 illustrates the operating principle of a biofilm in sewage treatment.

In an aerobic reactor, oxygen is consumed as it penetrates the biofilm, until anoxic or anaerobic conditions are reached. Therefore, an external layer with oxygen and an internal layer deprived of oxygen may be found. DO is the determining factor in the establishment of the layers. Nitrate reduction will occur in the anoxic layer. In anaerobic conditions, there will be the formation of organic acids and a reduction of sulphates. This coexistence between aerobic, anoxic and anaerobic conditions is an important characteristic of biofilm systems (Iwai and Kitao, 1994).

The biofilm formation process can be understood as occurring at three stages (Iwai and Kitao, 1994). Table 1.5 and Figure 1.10 present the main characteristics of these three stages associated with the thickness of the biofilm.

Table 1.5. Stages in the formation of the biofilm

Biofilm thickness	Characteristics
Thin	• The film is thin and frequently does not cover all the surface of the support medium • The bacterial growth follows a logarithmic rate • All the microorganisms grow under the same conditions, and the growth is similar to that of the dispersed biomass
Intermediate	• The thickness of the film becomes greater • The bacterial growth rate becomes constant • The thickness of the active layer stays unaltered, independently of the increase in the total thickness of the biofilm • If the supply of organic matter is limited, the microorganisms adopt a metabolism sufficient only for maintenance, but without growth • If the supply of organic matter is lower that the requirements for maintenance, the film thickness decreases
High	• The thickness of the biofilm reaches a very high level • The microbial growth is counteracted by the decay of the organisms, by the uptake by other organisms and by shearing stress • Parts of the biofilm can be dislodged from the support medium • If the biofilm continues to grow without being dislodged from the support medium, clogging will take place

Source: based on Iwai and Kitao (1994)

SUBSTRATE (S) CONCENTRATION GRADIENT IN A BIOFILM

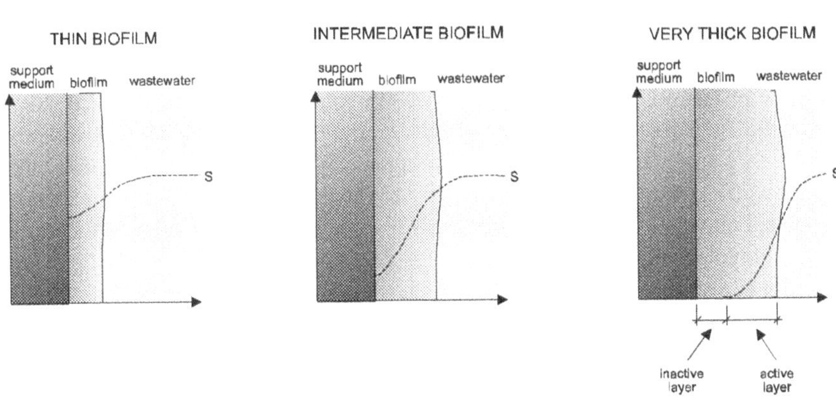

Figure 1.10. Concentration gradients of the substrate (S) in biofilms of different thicknesses (adapted from Lubberding, 1995)

When analysing *dispersed growth* and *attached growth* in sewage treatment, the comparison between the hydraulic detention time and the cell doubling time is an aspect of great importance. In dispersed growth systems, in order to have microbial population growth, the hydraulic detention time (average time that a water molecule stays in the system) has to be greater than the doubling time of the microorganisms, that is, the time necessary to generate new cells. If the hydraulic detention time is less than the cell doubling time, the bacteria are *"washed"* out of the system. This is a limiting factor for sizing the biological reactors, considering that reactor volume and detention time are directly related (detention time = volume/flow).

In the case of systems with biofilms, the hydraulic detention times can be less than the cell doubling time, without cell wash-out occurring, because of the fact that the bacteria are attached to a support medium. Consequently, it is possible to adopt lower volumes for the reactor.

In the comparison between dispersed-growth and attached-growth systems, there are the following aspects relative to attached-growth systems (Iwai and Kitao, 1994; Lubberding, 1995):

- The reactor can be operated with a hydraulic detention time lower than the cell doubling time.
- The concentration of *active* biomass can be higher than for dispersed-growth systems (see explanation below).
- The substrate removal rate can be higher than for dispersed-growth systems (see explanation below).
- The coexistence between aerobic and anaerobic microorganisms is greater than in the dispersed-growth systems because the thickness of the biofilm is usually greater than the diameter of the biological floc.
- The cells are fixed in the solid phase, while the substrate is in the liquid phase. This separation reduces the need or the requirements for the subsequent clarification stage.
- The microorganisms are continually reused. In the dispersed-growth system, reutilisation can only be implemented through recirculation of the biomass.
- If the biofilm thickness is high, there can be limitations for the diffusion of the substrate into the biofilm.

The potential difference between the activity of the dispersed and attached biomass and the consequent substrate removal rate can be explained as follows (Lubberding, 1995). The **dispersed biomass** has a density close to the sewage and moves itself in practically the same direction and velocity of the sewage inside the reactor. As a result, the biomass stays exposed to the same fraction of liquid for a larger period, leading to a low substrate concentration in the neighbourhood of the cell. With the low substrate concentrations, the bacterial activity and the substrate removal rate are also lower. Only at a certain distance from the cell is the substrate concentration higher. Considering the dependence between the substrate concentration and the microbial activity, the importance represented by the mixing level in the reactor becomes evident.

In the **attached biomass** systems, the density of the support medium together with the biomass is very different from the density of the liquid in the reactor, allowing the occurrence of velocity gradients between the liquid and the external layer of the biofilm. As a result, the cells are continually exposed to new substrates, potentially increasing their activity. However, if the biofilm thickness is very high, the substrate consumption along the biofilm could be such, that the internal layers have substrate deficiencies, which reduce their activity. In these conditions, the attachment with the support medium reduces and the biomass can be dislodged from the support medium.

2

Reaction kinetics and reactor hydraulics

2.1 INTRODUCTION

All biological wastewater treatment processes occur in a volume defined by specific physical boundaries. This volume is commonly called a *reactor*. The modifications in the composition and concentration of the constituents during the residence time of the wastewater in the reactor are essential items in sewage treatment. These changes are caused by:

- *hydraulic transportation of the materials in the reactor (input and output);*
- *reactions that occur in the reactor (production and consumption).*

The knowledge of the two components, which characterise the so-called *mass balance* around the reactor, is fundamental in the design and operation of wastewater treatment plants. Finally, the manner and efficiency in which these changes take place depend on the type and configuration of the reactor, which is dealt with in the study called *reactor hydraulics*.

The present chapter covers the following main topics:

- reaction kinetics
- mass balance
- reactor hydraulics

2.2 REACTION KINETICS

2.2.1 Reaction types

Most reactions that take place in sewage treatment are slow and the consideration of their kinetics is important. The *reaction rate* r is the term used to represent the disappearance or formation of a constituent or chemical species. The relation between the *reaction rate, concentration of the reagent* and the *order of reaction* is given by the expression:

$$r = kC^n \tag{2.1}$$

where:

　r = reaction rate $(ML^{-3}T^{-1})$
　k = reaction constant (T^{-1})
　C = reagent concentration (ML^{-3})
　n = reaction order

For different values of n, there are the following types of reactions:

- n = 0 *zero-order reaction*
- n = 1 *first-order reaction*
- n = 2 *second-order reaction*

When there is more than one reagent involved, the calculation of the reaction rate must take into consideration their concentrations. If there are two chemicals with concentrations A and B, the rate is:

$$r = kA^n B^m \tag{2.2}$$

The global reaction rate is defined as (m + n). For example, if a global reaction rate was determined as being $r = kA^2B$, the reaction is considered second order with relation to reagent A and first order with relation to reagent B. The global reaction rate is of third order (Tchobanoglous and Schroeder, 1985).

If the logarithm is applied on both sides of Equation 2.1 for a reaction with only one reagent, the following equation is obtained:

$$\log r = \log k + n \log C \tag{2.3}$$

The visualisation of the above relation for different values of n is presented in Figure 2.1. The interpretation of Figure 2.1 is:

- The *zero-order reaction* results in a horizontal line. The reaction rate is independent of the reagent concentration, that is, it is the same independently of the reagent concentration.
- The *first-order reaction* has a reaction rate directly proportional to the reagent concentration.

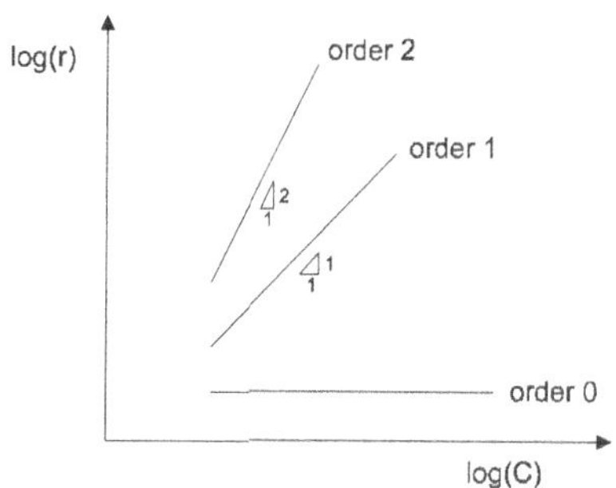

Figure 2.1.   Determination of the reaction order on a logarithmic scale (adapted from Benefield and Randall, 1980)

- The *second-order reaction* has a reaction rate proportional to the square of the reagent concentration.

The most frequent reaction orders found in sewage treatment are zero order and first order. Second order reactions may occur with some specific industrial wastewaters. The reaction order does not necessarily need to be an integer, and the laboratory determination of the decomposition rates of certain industrial constituents can lead to intermediate orders. Besides these reactions with constant order, there is another type of reaction, which is widely used in the area of wastewater treatment, called *saturation reaction*. In summary, the following reactions are going to be analysed in detail:

- *zero-order reaction*
- *first-order reaction*
- *saturation reaction.*

2.2.2 Zero-order reactions

Zero-order reactions are those in which the *reaction rate is independent of the reagent concentration*. In these conditions, the *rate of change of the reagent concentration (C) is* **constant**. This comment assumes that the reaction occurs in a batch reactor (see Item 2.4), in which there is no addition or withdrawal of the reagent during the reaction. In the case of a reagent that is disappearing in the reactor (for example, through decomposition mechanisms), the rate of change is given by Equation 2.4. The *minus* sign in the term on the right-hand side of the equation indicates *removal* of the reagent, whereas a *plus* sign would indicate *production* of the constituent.

$$\frac{dC}{dt} = -K.C^0 \qquad\qquad (2.4)$$

or

$$\boxed{\frac{dC}{dt} = -K}$$

(2.5)

The development of the rate of change (dC/dt) with time according to Equation 2.5 can be seen in Figure 2.2.a. It is seen that the rate is constant with time.

The integration of Equation 2.5 with $C = C_o$ at $t = 0$ leads to:

$$\boxed{C = C_o - K.t}$$

(2.6)

This equation can be visualised in Figure 2.2.b.

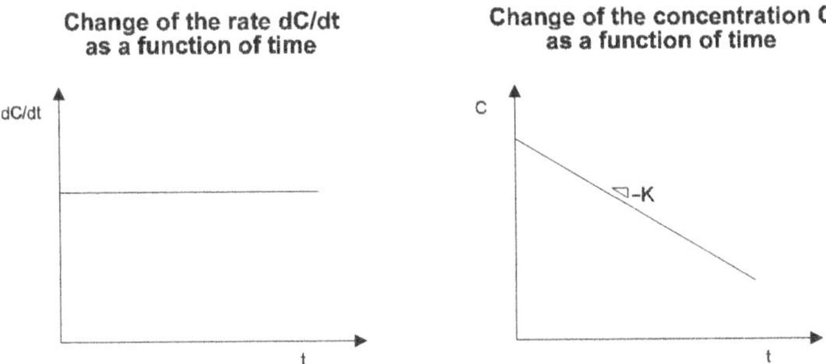

Figure 2.2.   Zero-order reactions. (a) Change of the reaction rate dC/dt with time. (b) Change of the concentration C with time.

2.2.3 First-order reactions

First-order reactions are those in which the *reaction rate is proportional to the concentration of the reagent*. Therefore, in a batch reactor, the rate of change of the reagent concentration C is proportional to the reagent concentration at a given time. Assuming a reaction in which the constituent is being removed, the associated equation is:

$$\frac{dC}{dt} = -K.C^1$$

(2.7)

or

$$\boxed{\frac{dC}{dt} = -K.C}$$

(2.8)

The development of the rate of change (dC/dt) with time according to Equation 2.8 is presented in Figure 2.3.a. It is noted that the rate decreases linearly with time.

Integrating Equation 2.8 with $C = C_o$ at $t = 0$ leads to:

$$\boxed{\ln C = \ln C_o - K.t}$$

(2.9)

or

$$\boxed{C = C_o.e^{-Kt}}$$

(2.10)

Equation 2.10 is plotted in Figure 2.3.b.

FIRST-ORDER REACTIONS

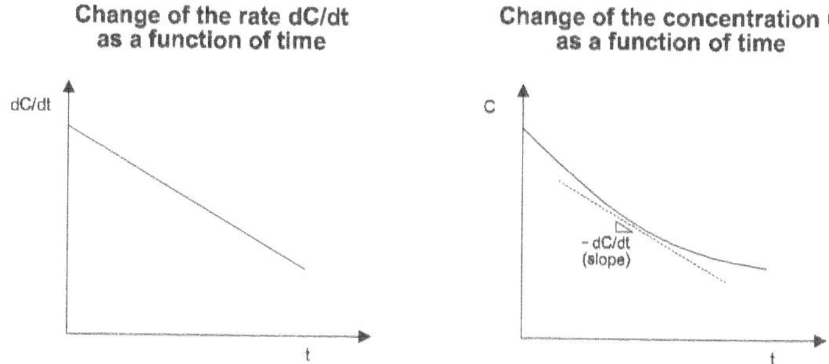

Figure 2.3.   First-order reactions. (a) Change of the reaction rate dC/dt with time. (b) Change of the concentration C with time.

Various reactions in sewage treatment follow first-order kinetics. The introduction of oxygen by artificial aeration is an example. Other examples are the removal of organic matter in some systems and the decay of pathogenic organisms. The biological stabilisation of the organic matter may be represented by pseudo-first-order reaction, as covered in Section 2.2.4. Although various components are involved, such as oxygen concentration, number of microorganisms and concentration of the organic matter, the rate can be proportional to the concentration of one constituent (organic matter, in this case), provided the others are in relative abundance (Arceivala, 1981). However, if the organic matter is available in low concentrations, the reaction occurs as a first-order reaction. This aspect is discussed in Section 2.2.4.

The global rate follows first-order kinetics in various complex processes. Many substances can individually show zero-order kinetics, but the complex substrates in which many of these substances are aggregated (e.g. domestic and industrial wastewater) can suffer a decay rate that follows a first-order reaction. Initially, when most of the components are being simultaneously removed (consumed), the global removal rate is high. After a certain time, however, the rate can be slower, when only the most hardly degradable constituents are still present. In this case, the global reaction rate may resemble a typical first-order reaction (Arceivala, 1981).

2.2.4 Saturation reactions

Another expression used to describe the rates involved in biological sewage treatment is based on enzymatic reactions, whose kinetics were proposed by Michaelis and Menten. Since bacterial decomposition involves a series of reactions catalysed by enzymes, the structure of the Michaelis–Menten expression can be used to describe the kinetics of bacterial growth and the decomposition reactions in sewage (Sawyer and McCarty, 1978) (see Chapter 3).

The reaction rate follows a hyperbolic form, in which the rate tends to a saturation value (Equation 2.11):

$$r = r_{max} \cdot \frac{S}{K_s + S} \qquad (2.11)$$

where

r = reaction rate $(ML^{-3}T^{-1})$
r_{max} = maximum reaction rate $(ML^{-3}T^{-1})$
S = concentration of the limiting substrate (ML^{-3})
K_s = half-saturation constant (ML^{-3})

Through Equation 2.11, it is seen that K_s is the concentration of the substrate in which the reaction rate r is equal to $r_{max}/2$. Equation 2.11 is illustrated in Figure 2.4.

Equation 2.11 is widely used in biological sewage treatment. Its large importance resides in its form that can approximately represent zero-order and first-order

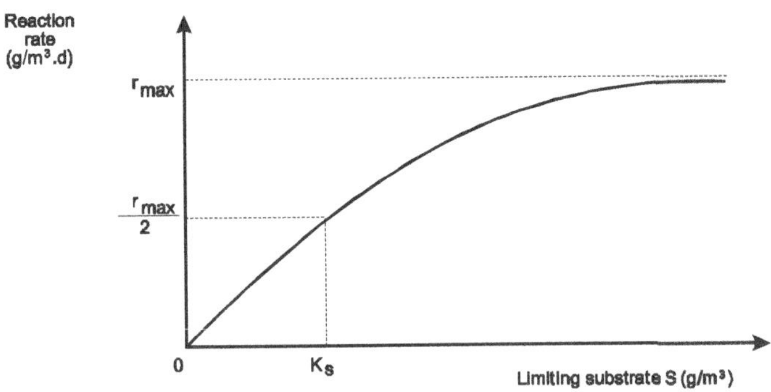

Figure 2.4. Graphical representation of the saturation reaction, according to Michaelis–Menten

kinetics (see Figure 2.5), as well as the transition between them. As mentioned before, at the start of a reaction of decomposition of substrate (organic matter), when the concentration is still high, there is no substrate limitation in the medium and the global removal rate approaches zero-order kinetic. When the substrate starts to be consumed, the reaction rate starts to decrease, characterising a transition region or mixed order. When the substrate concentration is very low, the reaction rate starts to be limited by the low availability of organic matter in the medium. In these conditions, the kinetics develops as in a first-order reaction. These situations occur as a function of the relative values of S and K_s, as described below.

- *Relative substrate concentration: **high***

$$\boxed{S >> K_s : approximately\ zero\text{-}order\ reaction}$$

When the substrate concentration is much higher than the value of K_s, K_s can be neglected in the denominator of Equation 2.11. The equation is thus reduced to:

$$r = r_{max} \tag{2.12}$$

In these conditions, the reaction rate r is constant and equal to the maximum rate r_{max}. The reaction follows zero-order kinetics, in which the reaction rate is independent of the substrate concentration. In domestic sewage treatment, such a situation tends to occur, for instance, at the head of a plug-flow reactor, where the substrate concentration is still high.

- *Relative substrate concentration: **low***

$$\boxed{S << K_s : approximately\ low\text{-}order\ reaction}$$

When the substrate concentration is much lower than the value of K_s, S can be ignored in the denominator of Equation 2.11. Consequently, the equation is reduced to:

$$r = r_{max} \cdot \frac{S}{K_S} \tag{2.13}$$

Since r_{max} and K_s are two constants, the term (r_{max}/K_s) is also a constant, and can be substituted by a new constant K. Thus, Equation 2.13 is reduced to:

$$r = K.S \tag{2.14}$$

In this situation, the reaction rate is proportional to the substrate concentration. The reaction follows *first-order kinetics*. Such a situation is typical in domestic sewage treatment, in complete-mix reactors, where the substrate

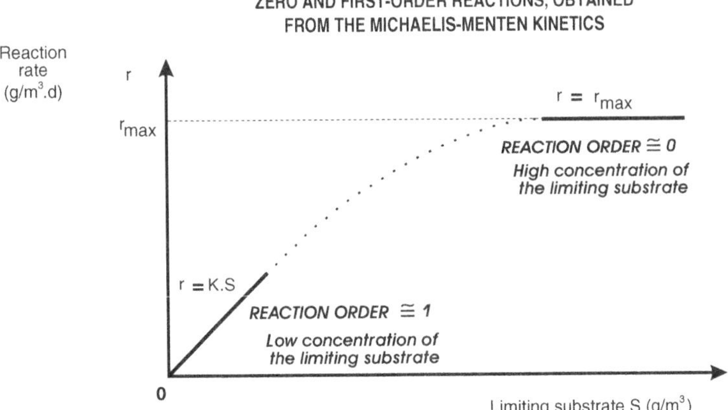

Figure 2.5. Michaelis–Menten kinetics. Two extreme conditions in the saturation reaction

concentration in the medium is low, due to the requirements of having low substrate levels in the effluent.

Figure 2.5 presents the two extreme situations, representatives of zero and first order-kinetics.

The form of the Michaelis–Menten equations is reanalysed in detail in Chapter 3 that treats the kinetics of bacterial growth and substrate removal. In these conditions, one has the so-called Monod kinetics. In Chapter 3, typical values of K_s for the treatment of domestic sewage are presented, and the conditions that lead to the predominance of the zero or first-order kinetics are re-evaluated.

2.2.5 Influence of the temperature

The rate of any chemical reaction increases with temperature, provided that this increase in temperature does not produce alterations in the reagents or in the catalyst. The biological reactions, within certain ranges, also present the same tendency to increase with temperature. However, there is an ideal temperature for the biological reactions, above which the rate decreases, possibly due to the destruction of enzymes at the higher temperatures (Sawyer and McCarty, 1978; Benefield and Randall, 1980).

A usual form to estimate the variation of the reaction rate as a function of temperature is through the formulation based on the van't Hoff-Arrhenius theory, which can be expressed as:

$$\boxed{\frac{K_{T_2}}{K_{T_1}} = \theta^{T_2 - T_1}} \tag{2.15}$$

where:

K_{T_2} = reaction coefficient for temperature 2
K_{T_1} = reaction coefficient for temperature 1
θ = temperature coefficient

Even thought θ is frequently treated as a constant, it can vary substantially, even inside a reduced temperature range (Sawyer and McCarty, 1978). The values of θ usually adopted for the corrections of various reactions involved in sewage treatment are presented in the relevant chapters of this book.

2.3 MASS BALANCE

2.3.1 Representative equations

Once the reaction rates of interest are known, their influence on the general mass balance of the constituent under analysis must be quantitatively evaluated. This is because the concentration of a certain constituent in a reactor (or in any place inside it) is a function, not only of the biochemical reactions, but also of the transport mechanisms (input and output) of the constituent. *Reactor* is the name given to tanks or generic volumes in which the chemical or biological reactions occur.

The mass balance is a quantitative description of all the materials that enter, leave and accumulate in a system with defined physical boundaries. The mass balance is based on the law of conservation of mass, that is, mass is neither created nor destroyed. The basic mass balance expression should be derived in a chosen volume, which can be either a tank or a reactor as a whole, or any volume element of them. In the mass balance, there are terms for (Tchobanoglous and Schroeder, 1985):

- *materials that enter*
- *materials that leave*
- *materials that are generated*
- *materials that are consumed*
- *materials that are accumulated in the selected volume*

In any selected volume (see Figure 2.6), the quantity of accumulated material must be equal to the quantity of material that enters, minus the quantity that leaves, plus the quantity that is generated, minus the quantity that is consumed. In linguistic terms, the mass balance can be expressed in the following general form.

$$\boxed{\text{Accumulation} = \text{Input} - \text{Output} + \text{Production} - \text{Consumption}} \quad (2.16)$$

Some authors prefer not to include in an explicit form the term relative to consumption, which must be expressed as a produced material, with a minus sign in the reaction rate. The convention adopted in this text is the one of Equation 2.16, which leads to a clearer understanding of the four main components involved in

MASS BALANCE IN A REACTOR

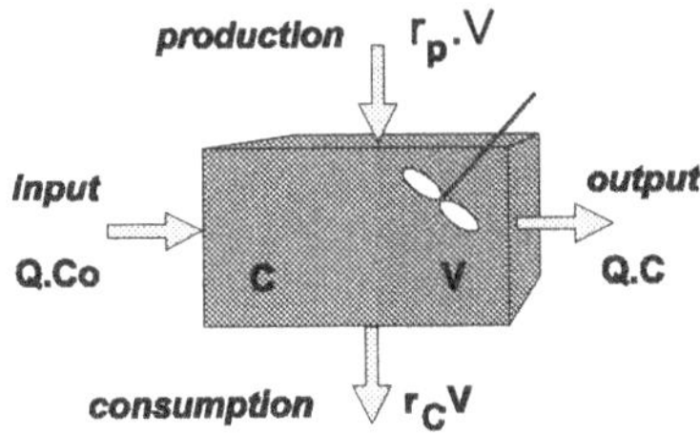

Figure 2.6.   Mass balance in a reactor

the mass balance. Therefore, care and coherence must be exercised with the signs
of each term when adopting one convention or another.

Mathematically the relation of Equation 2.16 can be expressed as:

$$\frac{d(C.V)}{dt} = Q.C_0 - Q.C + r_p.V - r_c.V \qquad (2.17)$$

where
  C = concentration of the constituent at a time t (ML^{-3})
  C_o = influent concentration of the constituent (ML^{-3})
  V = volume of the reactor (completely mixed) or volume element of any
      reactor (L^3)
  Q = flow (L^3T^{-1})
  t = time (T)
  r_p = reaction rate of production of the constituent $(ML^{-3}T^{-1})$
  r_c = reaction rate of consumption of the constituent consumed $(ML^{-3}T^{-1})$

Equation 2.17 can be expressed in the following alternative form, in which the
left-hand term has been expanded:

$$C.\frac{dV}{dt} + V.\frac{dC}{dt} = Q.C_0 - Q.C + r_p.V - r_c.V \qquad (2.18)$$

The volume in biological reactors can usually be considered as fixed (dV/dt=0),
making the first term on the left-hand side disappear. This leads to the simplified
and more usual form of the mass balance, presented in Equation 2.19. Since in
this equation the only dimension is time, the formulation is of a *ordinary differential equation*, in which the analytical solution (or numeric computation) is much
simpler. However, it must be emphasised that the mass balance in other systems,
such as, for instance, the sludge volume in secondary sedimentation tanks in activated sludge systems, also implies variations in volume (besides concentration
variations). In this particular case, there are two dimensions (time and space), which

lead to a *partial differential equation*. The solution of these equations demands a greater mathematical sophistication. However, for completely-mixed biological reactors (fixed volume), the more usual mass balance, expressed in Equation 2.19, is used.

$$V.\frac{dC}{dt} = Q.C_0 - Q.C + r_p.V - r_c.V \qquad (2.19)$$

In the preparation of a mass balance, the following steps must be followed (Tchobanoglous and Schroeder, 1985):

* Prepare a simplified schematic or flowsheet of the system or process for which the mass balance will be prepared.
* Draw the system boundaries, to define where the mass balance will be applied.
* List all the relevant data that will be used in the preparation of the mass balance in the schematic or flowsheet.
* List all the chemical or biological reaction equations that are judged to represent the process.
* Select a convenient basis on which the numerical calculations will be done.

2.3.2 Steady state and dynamic state

The mathematical model of the system can be structured for two distinct conditions:

* *Steady state*
* *Dynamic state*

The **steady state** is the one in which there are no accumulations of the constituent in the system (or in the volume being analysed). Thus, $dC/dt = 0$, that is, *the concentration of the constituent is constant*. In the steady state, the input and output flows and concentrations are constant. There is a perfect equilibrium between the positive and the negative terms in the mass balance, which, when summed, lead to a zero value. In the *design* of wastewater treatment plants, it is more usual to use the simplified steady state equations. Under these conditions, in which $dC/dt = 0$, the mass balance is given by Equation 2.20.

$$0 = Q.C_0 - Q.C + r_p.V - r_c.V \qquad (2.20)$$

The **dynamic state** is the one in which there are mass accumulations of the constituent in the system. Hence, $dC/dt \neq 0$. *The concentration of the constituent in the system is therefore variable with time* and can increase or decrease, depending on the balance between the positive and negative terms. Usually in a treatment plant, the input flow and/or the input concentration are variable, besides the possibility

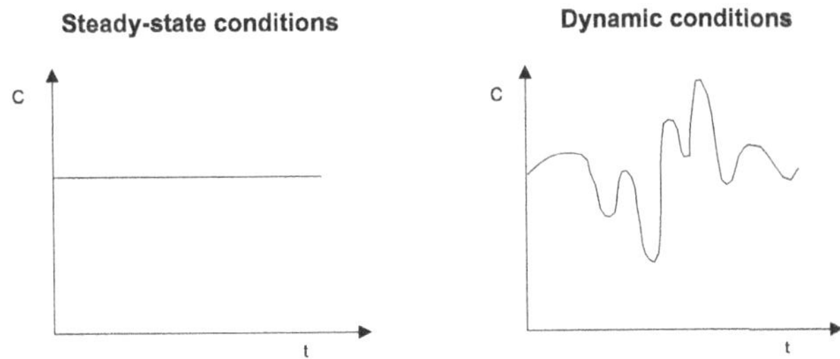

Figure 2.7.   Steady and dynamic conditions. Profile of the concentration of the constituent with time.

of having other external stimulus to the system (temperature changes) that cause a *transient* in the concentration of the constituent. For this reason, dynamic conditions are the ones really prevailing in actual sewage treatment plants. The dynamic models are based on the generalised mass balance equation (Equation 2.19). For the *operational control* of a treatment plant, dynamic models are more adequate, due to the frequent variation of the external and internal conditions of the system. The dynamic models can be also used for design, principally for evaluating the impact of variable influent loads on the performance of the plant. The dynamic models have been less used due to the larger complexity involved in the solution of the equations and the greater requirements of values for model coefficients and variables. However, the trend of using more computers and numerical integration routines commercially available has contributed to a greater use of dynamic models. It must be emphasised that the steady state is only a particular case of the dynamic state.

Figure 2.7 illustrates the concept of the steady and dynamic states, through the representation of the variation of the concentration of the constituent with time.

2.4 REACTOR HYDRAULICS

2.4.1 Introduction

After the reaction rates are known (Section 2.2) and the mass balances have been established (Section 2.3), in order to calculate the concentration of the constituent in the reactor it is necessary to define the hydraulic model to be attributed to it.

The hydraulic model of the reactor is a function of the type of flow and the mixing pattern in the unit. The mixing pattern depends on the physical geometry of the reactor, the quantity of energy introduced per unit volume, the size or scale of the unit and other factors.

In terms of flow, there are the following two conditions:

- *Intermittent flow* (batch): discontinuous input and/or output
- *Continuous flow*: continuous input and output

In terms of the mixing pattern, there are two basic idealised hydraulic models, which define an envelope inside which the other patterns are found. These are the **plug-flow** and the **complete-mix** reactors, which lead to the following main alternatives:

- *plug flow reactor*
- *complete-mix reactor*
- *dispersed flow*
- *reactors in series and/or in parallel.*

The main types of reactors used in sewage treatment are presented in Table 2.1 (Tchobanoglous and Schroeder, 1985; Metcalf & Eddy, 1991), and their operational characteristics are summarised in Table 2.2.

Table 2.1. Characteristics of the most frequently used reactors and hydraulic models in sewage treatment

Hydraulic model	Schematics	Characteristics
Batch reactors		A batch reactor is the one in which there is no flow entering or leaving. The reactor contents are completely mixed. All elements are exposed to the treatment for a time equal to the substrate residence time in the reactor. Consequently, the batch reactor behaves like a discrete volume of a plug-flow reactor. The BOD test bottle is an example of a batch reactor. One of the variants of the activated sludge process is the sequencing batch reactors.
Plug flow		The fluid particles enter the tank continuously in one extremity, pass through the reactor, and are then discharged at the other end, in the same sequence in which they entered the reactor. The fluid particles move as a piston, without any longitudinal mixing. The particles maintain their identity and stay in the tank for a period equal to the theoretical hydraulic detention time. This type of flow is reproduced in long tanks with a large length-to-breadth ratio, in which longitudinal dispersion is minimal. These reactors are also called tubular reactors. Plug-flow reactors are idealised reactors, since complete absence of longitudinal dispersion is difficult to obtain in practice.

(Continued)

Table 2.1 (*Continued*)

Hydraulic model	Schematics	Characteristics
Complete mix		The particles that enter the tank are immediately dispersed in all the reactor body. The input and output flows are continuous. The fluid particles leave the tank in proportion to their statistical population. Complete mix can be obtained in circular or square tanks in which the tank's contents are continuously and uniformly distributed. Complete-mix reactors are also known as CSTR or CFSTR (continuous-flow stirred tank reactors). Complete-mix reactors are idealised reactors, since total and identical dispersion is difficult to obtain in practice.
Dispersed flow		Dispersed or arbitrary flow is obtained in any reactor with an intermediate degree of mixing between the two idealised extremes of plug flow and complete mix. In reality, most reactors present dispersed-flow conditions. However, due to the greater difficulty in their modelling, the flow pattern is frequently represented by one of the two idealised hydraulic models. The input and output flows are continuous.
Complete-mix reactors in series		Complete-mix reactors in series are used to model the hydraulic regime that exists between the idealised plug flow and complete mix regimes. If the series is composed of only one reactor, the system reproduces a complete-mix reactor. If the system has an infinite number of reactors in series, plug flow is reproduced. Input and output flows are continuous. Reactors in series are also commonly found in maturation ponds.
Packed-bed reactors		These reactors are filled with some type of packing medium, such as stone, plastic, ceramic and others. Regarding the flow and saturation, these reactors can be submerged, with the pores saturated (anaerobic filters and submerged aerated filters) or with intermittent dosing, with non-saturated pores (trickling filters). The flow can be upward or downward.

Table 2.2. Operational characteristics of the main reactor systems (assuming steady-state conditions)

Reactor type	Schematics	Continuous flow	Variation of the composition with time (in a given position in the reactor)	Variation of the composition with the position in the reactor (at a given time)	Number of equivalent complete mix reactors	Typical length / breadth ratio
Batch reactor		No	Yes	No	–	≈ 1
Plug flow		Yes	No	Yes	∞	$>> 1$
Complete mix		Yes	No	No	1	≈ 1

2.4.2 Ideal plug-flow reactor

The ideal plug flow is the one in which each fluid element leaves the tank in the same order of entrance. No element anticipates or delays another in the journey. The flow occurs as pistons moving from upstream to downstream, without mixing between the pistons and without dispersion. Consequently, each element is exposed to treatment for the same period of time (as in a batch reactor), which is equal to the theoretical hydraulic detention time (Arceivala, 1981).

Figure 2.8 presents a summary of the concentration profiles with time and position in an ideal plug-flow reactor submitted to constant influent flow and concentration (steady-state conditions). If the influent load is varied (dynamic conditions), the derivation of the formulas for the plug-flow reactor is more complicated than for complete mix. This is because the concentration in the plug flow varies with time and space in the reactor, while in complete mix the variation is only with time (same concentration in any position of the reactor). That is why complete-mix reactors in series are frequently used to simulate a plug-flow reactor under dynamic (time-varying) conditions.

If the influent (input) concentration is constant, the effluent concentration (output) also remains constant with time. The concentration profile in the tank and, therefore, the effluent concentration, depend on the type and reaction rate of the constituent. Table 2.3 summarises the main intervening equations.

The following generalisations can be made for an ideal plug-flow reactor under steady state conditions:

- *Conservative substances*: the effluent concentration is equal to the influent concentration.
- *Biodegradable substances with zero-order reaction*: the removal rate is constant from the inlet to the outlet end of the reactor.
- *Biodegradable substances with first-order reaction*: along the reactor, the substrate removal coefficient (K) is constant, but the concentration decreases gradually while the wastewater flows along the reactor. At the inlet end of the reactor, the concentration is high, which causes the removal rate to be also high (in first-order reactions the removal rate is proportional to

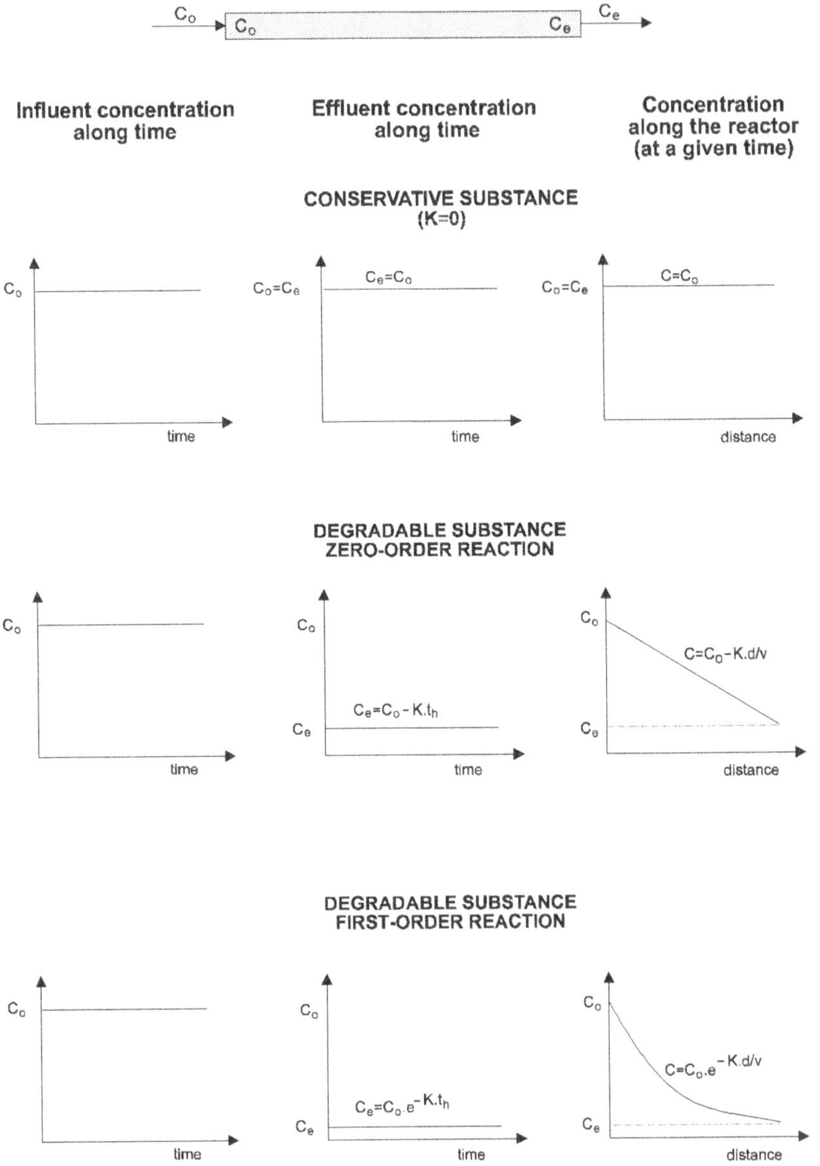

Figure 2.8. Concentration profiles – Ideal plug-flow reactor under steady-state conditions. Nomenclature: C = concentration at a given time; C_o = influent concentration; C_e = effluent concentration; K = reaction coefficient; t_h = hydraulic detention time; d = distance (length of the reactor); v = horizontal velocity. In this figure, time represents the operational time, and not the travel time along the reactor.

Table 2.3. Ideal plug-flow reactor. Steady-state conditions. Equations for the calculation of the concentration along the tank and the effluent concentration

Reaction	Concentration along the reactor (at a given time)	Effluent concentration
Conservative substance ($r_c = 0$)	$C = C_0$	$C_e = C_0$
Biodegradable substance (zero order reaction; $r_c = K$)	$C = C_0 - K.d/v$	$C_e = C_0 - K.t_h$
Biodegradable substance (first order reaction; $r_c = K.C$)	$C = C_0.e^{-K.d/v}$	$C_e = C_0.e^{-K.t_h}$

C = concentration at a given point in the reactor (g/m^3)
C_0 = influent concentration (g/m^3)
K = reaction coefficient (d^{-1})
d = distance along the tank (m)
v = horizontal velocity (m/d)
t_h = hydraulic detention time (= volume/flow) (d)

the concentration). At the outlet end of the reactor, the concentration is reduced and, consequently, the removal rate is low, that is, more time is required to reduce a unit value of the concentration.

- *First-order or higher reactions*: the plug flow is more efficient than the complete-mix reactor.

Example 2.1

A reactor with predominantly longitudinal dimensions has a volume of 3,000 m^3. The influent has the following characteristics: flow = 600 m^3/d; substrate concentration = 200 g/m^3.

Calculate the concentration profile along the reactor (assuming an ideal plug-flow reactor under steady state) in the following conditions:

- conservative substance ($K = 0$)
- biodegradable substance with first-order removal ($K = 0.40 \text{ d}^{-1}$)

Solution:

a) Hydraulic detention time

The hydraulic detention time (t_h) is given by:

$$t_h = \frac{V}{Q} = \frac{3000 \text{ m}^3}{600 \text{ m}^3/d} = 5 \text{ d}$$

The travel distance is proportional to the time spent for the piston to flow. The total distance is covered when the hydraulic detention time is reached.

Example 2.1 (Continued)

b) Conservative substance

The application of the formula $C = C_o e^{-K.t}$ for steady state, with $K = 0$, for various values of t, leads to:

$$C = 200.e^{-0xt}$$

Travel time (d)	Distance / total length	Concentration along the tank (g/m^3)
0	0.0	200
1	0.2	200
2	0.4	200
3	0.6	200
4	0.8	200
5	1.0	200

The same values can be obtained through the direct application of the formula $C = C_o$ (Table 2.3) for the conservative substances.

The effluent concentration is the concentration at the end of the hydraulic detention time ($t_h = 5$ d), that is, 200 g/m^3. The same value can be obtained through the direct application of the formula $C_e = C_o$ (Table 2.3).

The profile of the concentration along the tank is plotted below.

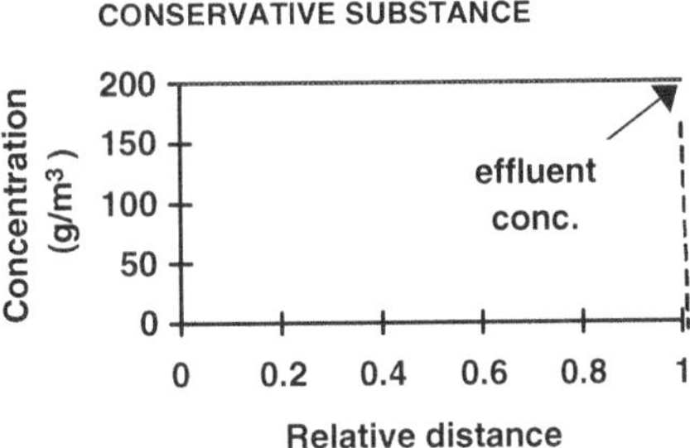

c) Biodegradable substance (with a first-order reaction)

The application of the formula $C = C_o e^{-Kt}$ (steady state) for various values of t leads to:

$$C = 200.e^{-0.40xt}$$

Example 2.1 (Continued)

Travel time (d)	Distance / total length	Concentration along the tank (g/m³)
0	0.0	200
1	0.2	134
2	0.4	90
3	0.6	60
4	0.8	41
5	1.0	27

The effluent concentration is the concentration at the end of the hydraulic detention time ($t_h = 5$ d), that is, 27 g/m³. The same value can be obtained through the direct application of the formula $C_e = C_0 e^{-K.t_h}$ (Table 2.3) for first-order reactions.

The concentration profile along the tank is plotted below.

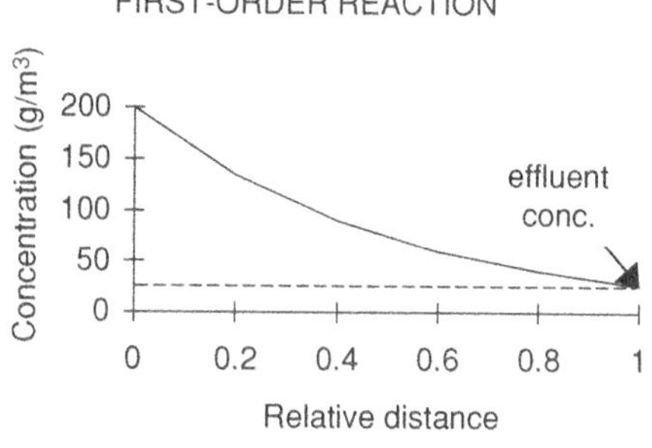

2.4.3 Ideal complete-mix reactor

The reactor with continuous flow and ideal complete mixing conditions is the one in which all of the elements that enter the reactor are instantaneously and totally dispersed. Thus, the reactor contents are homogeneous, that is, the concentration of any component is the same at any point in the tank. As a result, the effluent concentration is the same as that at any point in the reactor.

The mass balance in the reactor is (see Equations 2.16 and 2.19):

$$\text{Accumulation} = \text{Input} - \text{Output} + \text{Production} - \text{Consumption} \qquad (2.21)$$

$$\boxed{V.\frac{dC}{dt} = Q.C_0 - Q.C + r_p.V - r_c.V} \qquad (2.22)$$

Table 2.4. Ideal complete-mix reactor. Steady-state conditions. Equations for the calculation of the concentration along the tank and the effluent concentration

Reaction	Concentration along the reactor (at a given time)	Effluent concentration
Conservative substance ($r_c = 0$)	$C = C_o$	$C_e = C_o$
Biodegradable substance (zero order reaction; $r_c = K$)	$C = C_o - K.t_h$	$C_e = C_o - K.t_h$
Biodegradable substance (first order reaction; $r_c = K.C$)	$C = C_o/(1 + K.t_h)$	$C_e = C_o/(1 + K.t_h)$

C = concentration at a given point in the reactor (g/m^3)
C_o = influent concentration (g/m^3)
K = reaction coefficient (d^{-1})
d = distance along the tank (m)
t_h = hydraulic detention time (= volume/flow) (d)

In the steady state there is no mass accumulation in the reactor, that is, $dC/dt = 0$. In this analysis there is no production of constituents, only consumption reactions. Therefore, $r_p = 0$. Dividing the remaining terms by Q, and knowing that $t = V/Q$, the following equation is obtained:

$$0 = C_0 - C - r_c.t \qquad (2.23)$$

With the rearrangement of Equation 2.23, concentration profiles along the complete-mix reactor and the effluent concentration under steady-state conditions can be calculated (Figure 2.9).

If the influent (input) concentration is constant, the effluent (output) concentration also remains constant with time. The effluent concentration depends on the type and reaction rate of the constituent. However, the concentration profile along the reactor depicts a constant concentration, which is in agreement with the assumption that in a complete-mix reactor the concentrations are the same at any point in the tank. Table 2.4 summarises the main equations.

In comparison with the plug-flow reactor, the effluent concentration is only different for reactions of first order (or higher). For such reaction orders, the complete-mix reactor is less efficient than the plug-flow reactor.

The following generalisations can be made for an ideal complete-mix reactor under steady-state conditions:

- *Conservative and biodegradable substances*: the concentration and the removal rate are the same at any point in the reactor. The effluent concentration is equal to the concentration at any point in the reactor.
- *Conservative substances*: the effluent concentration is equal to the influent concentration.
- *Biodegradable substances with zero-order reaction*: the effluent concentration is equal to the effluent concentrations of a plug-flow reactor with the same detention time (the removal rate is independent of the local substrate concentration).

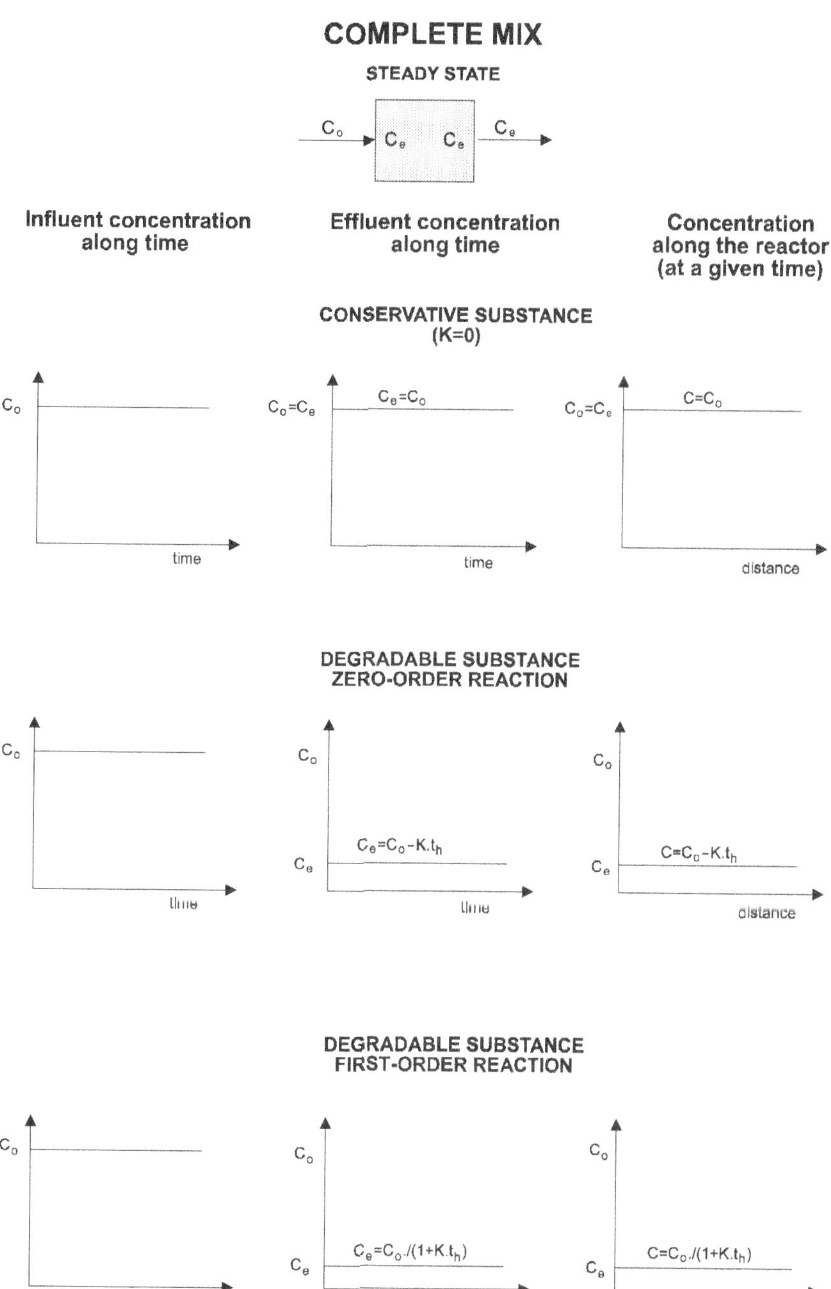

Figure 2.9.   Concentration profiles – Ideal complete mix reactor under steady-state conditions. Nomenclature: C = concentration at a given time; C_o = influent concentration; C_e = effluent concentration; K = reaction coefficient; t_h = hydraulic detention time. In this figure, time represents the operational time, and not the travel time along the reactor.

- *Biodegradable substances with first-order reactions or higher*: the complete-mix reactor is less efficient than the plug-flow reactor. Considering (a) that the removal rate is a function of the local concentration in first or higher-order reactions and (b) that the concentration at a complete-mix reactor is lower than the *average* concentration along a plug-flow reactor, then the efficiency of the complete-mix reactor is lower than that of the plug-flow reactor.

Example 2.2

A reactor of an approximately square shape and good mixing conditions has the same volume as the reactor in Example 2.1 (3,000 m^3). The influent also has the same characteristics of the referred example (flow = 600 m^3/d; influent substrate concentration = 200 g/m^3).

Calculate the concentration profile along the reactor (assuming an ideal complete-mix reactor under steady state) in the following conditions:

- Conservative substance (K = 0)
- Biodegradable substance with first-order removal (K = 0.40 d^{-1})

Solution:

a) Hydraulic detention time

The hydraulic detention time is the same calculated in Example 2.1 that is, t_h = 5 days.

b) Conservative substance

In a complete-mix reactor, the concentration is the same at any point. For a conservative substance, $C = C_o$ (Table 2.4). Hence, for any distance, the concentration is:

$$C = 200 \text{ g/m}^3$$

The effluent concentration is also equal to 200 g/m^3. This value is equal to that calculated for the ideal plug-flow reactor.

The concentration profile along the tank is plotted below.

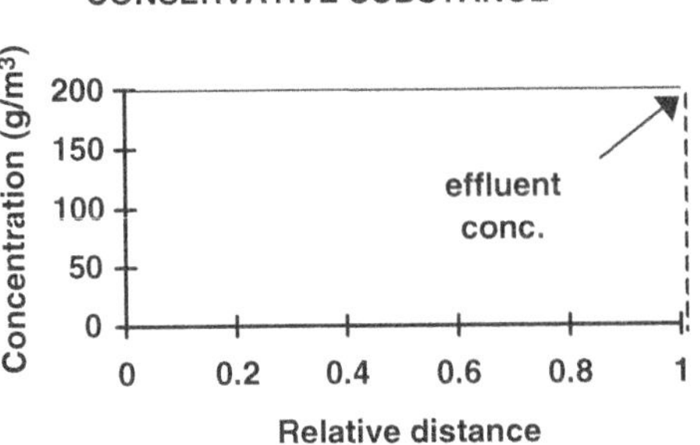

Example 2.2 (Continued)

c)   Biodegradable substance (with a first-order reaction)

At any point in the reactor, the concentration is given by:

$$C = \frac{C_o}{1 + K.t_h} = \frac{200}{1 + 0.40 \times 5} = 67 \text{ g/m}^3$$

The effluent concentration is also equal to 67 g/m³. This value is higher than the value calculated for the plug-flow reactor in Example 2.1 (27 g/m³), illustrating the fact that a complete-mix reactor is less efficient than a plug-flow reactor, for the same detention time.

The concentration profile along the tank is plotted below.

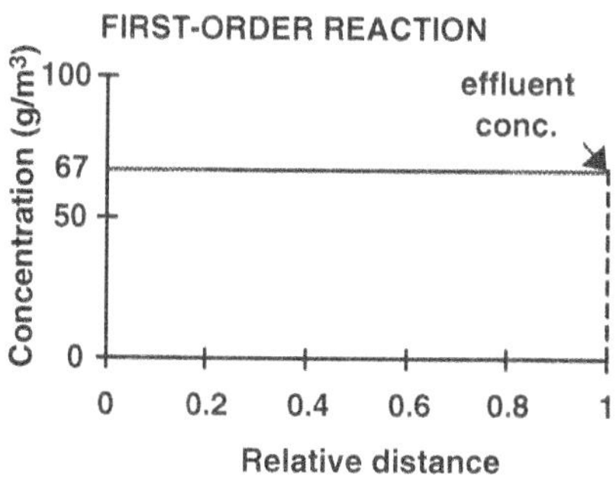

2.4.4 Cells in series

Another widely used hydraulic model is the complete-mix reactor in series, or cells in series. This system can occur in practice, such as in maturation ponds or activated sludge reactors with internal divisions, or it can be used as a theoretical model to represent intermediate hydraulic conditions between the complete-mix and the plug-flow reactor. When the total volume is distributed in *only one cell*, the system behaves like a conventional *complete-mix* reactor. Conversely, when the total volume is distributed in an *infinite number of cells*, the system reproduces *plug flow*. An *intermediate number of cells* simulates *dispersed flow*, with the system approaching the behaviour of complete mix or plug flow depending on the number of subdivisions adopted. When few cells are considered, the system tends to complete mix. On the other hand, when the system is subdivided into a larger number of cells it tends to plug flow.

Figure 2.10 presents the schematics of the two possible arrangements of cells in series, the first with cells of the same volume and the second with different volume cells.

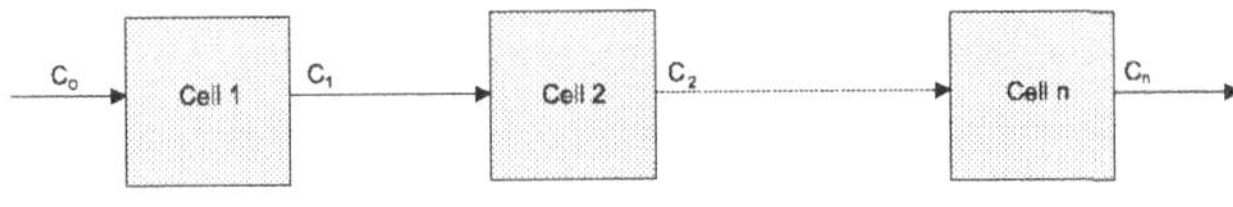

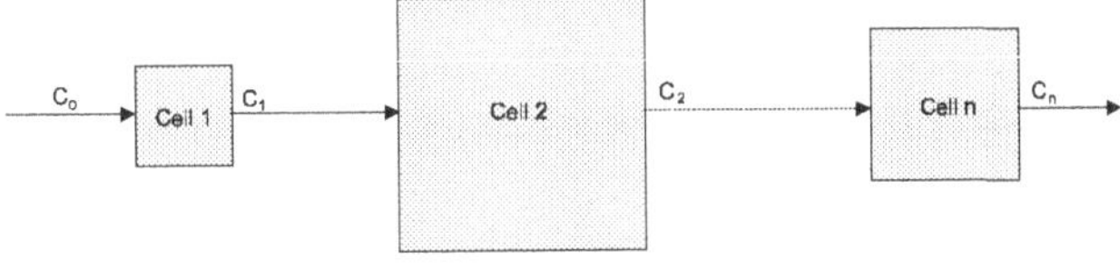

Figure 2.10.   Schematic arrangement of cells in series. (a) Equal cells; (b) different cells.

The effluent concentration from each cell is given by the same formulas for complete mix. Thus, there are three possible cases, depending on the removal rate:

- *Conservative substances*

Since there is no removal of conservative substances, the effluent from each cell is equal to its influent, which is also equal to the overall influent (in the steady state). Thus, the final effluent is given by:

$$C_e = C_o \qquad (2.24)$$

- *Biodegradable substances (zero-order removal)*

In zero-order reactions, the formula for a single cell is $C_e = C_o - K.t$. The effluent of the first cell is, therefore:

$$C_{e1} = C_o - K.t_1$$

where:
   C_{e1} = effluent concentration from the first cell
   t_1 = hydraulic detention time in the first cell

The effluent from the first cell is the influent to the second cell. Hence:

$$C_{e2} = C_{e1} - K.t_2 = C_o - K.t_1 - K.t_2$$

where:
   C_{e2} = effluent concentration from the second cell
   t_2 = hydraulic detention time in the second cell

For a system of n cells:

$$C_e = C_o - K.t_1 - K.t_2 - \cdots - K.t_n$$

$$C_e = C_o - K.(t_1 + t_2 + \cdots + t_n)$$

$$\boxed{C_e = C_o - K.t_h} \tag{2.25}$$

where:

C_e = final effluent concentration

t_h = hydraulic detention time in the system (summing up the volume of all cells) = $t_1 + t_2 + \cdots + t_n$

It can be observed from Equation 2.25 that the final effluent from a system of n cells in series with a *zero-order reaction* is equal to that from a one-cell complete-mix reactor (with a volume equal to the total volume of all the cells). Additionally, it must be noted that this final effluent is also equal to the effluent from a plug-flow reactor. This is as expected, considering that in zero-order reactions, the removal rate is independent of the concentration. Therefore, the three reactor systems behave in an identical manner.

- *Biodegradable substances (first-order removal)*

In the case of first-order reactions, the formula for a single cell is $C_e = C_o/(1 + Kt)$. Thus, the effluent from the first cell is:

$$C_{e1} = \frac{C_o}{(1 + K.t_1)}$$

The effluent from the first cell is the influent to the second cell. Hence:

$$C_{e2} = \frac{C_1}{(1 + K.t_2)} = \frac{C_0}{(1 + K.t_1).(1 + K.t_2)}$$

Generalising for n cells:

$$\boxed{C_{en} = \frac{C_0}{(1 + K.t_1).(1 + K.t_2)\ldots(1 + K.t_n)}} \tag{2.26}$$

If all the cells have the *same volume* (and, as a consequence, the same hydraulic detention time), Equation 2.26 is simplified to:

$$\boxed{C_e = \frac{C_o}{(1 + K.t_1)^n} = \frac{C_0}{\left(1 + K.\frac{t_h}{n}\right)^n}} \tag{2.27}$$

Table 2.5. Complete-mix cells in series. Steady-state conditions. Equations for the determination of the final effluent concentration

Reaction	Cells with different sizes	Cells with equal sizes
Conservative substance ($r_c = 0$)	$C_e = C_o$	$C_e = C_o$
Biodegradable substance (zero-order reaction; $r_c = K$)	$C_e = C_o - K.t_h$	$C_e = C_o - K.t_h$
Biodegradable substance (first-order reaction; $r_c = K.C$)	$C_e = C_o/[(1 + K.t_1) \times (1 + K.t_2) \times \ldots \times (1 + K.t_n)]$	$C_e = C_o/(1 + K.t_1)^n$ $= 1/(1 + K.t_h/n)^n$

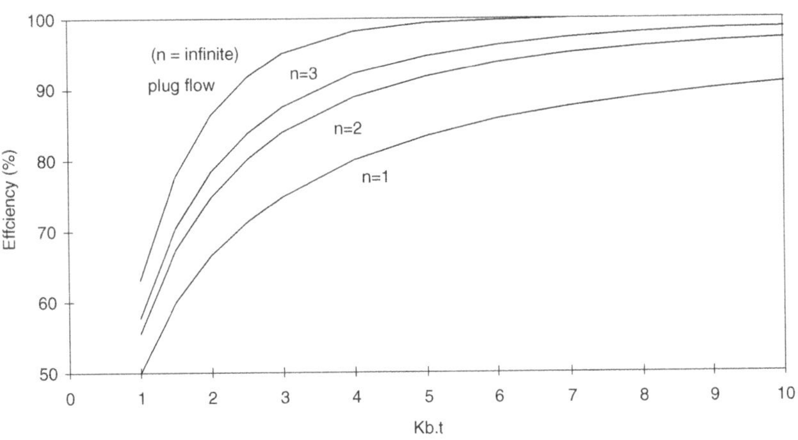

Figure 2.11.   Removal efficiencies for first-order kinetics in a system composed of CSTR cells in series, as a function of the dimensionless product K.t

where:

C_e = final effluent concentration (g/m^3)

K = reaction coefficient (d^{-1})

t_1 = hydraulic detention time for only one cell (d)

t_h = total hydraulic detention time in the system (summing up the volume of all the cells) (d)

n = number of cells, all having the same volume

Table 2.5 presents a summary of the formulas for the calculation of the effluent concentration from a system composed of n cells in series.

Fig. 2.11 presents the removal efficiencies for first-order kinetics in a system composed of equal-sized CSTR cells in series, as a function of the dimensionless product K.t. The great influence of the number of cells is clearly seen.

In many practical applications, it should be taken into account that the reaction coefficient K may vary from cell to cell. For instance, the first cell, receiving more highly biodegradable substance may have a higher K value than the subsequent cells which receive a less biodegradable influent (because the compounds more easily biodegradable have been removed in the first cell).

Example 2.3

A system with three equal cells in series has the same *total* volume of the reactor in Example 2.1 (3,000 m^3). The influent also has the same characteristics from the referred example (flow = 600 m^3/d; influent substrate concentration = 200 g/m^3).

Calculate the concentration profile along the system. Assume that each cell is an ideal complete-mix reactor in a steady state and that the substance is biodegradable with first-order removal (K = 0.40 d^{-1}).

Solution:

a) Hydraulic detention time in each cell

The hydraulic detention time in each cell is equal to the total detention time divided by the number of cells, that is:

$$t_1 = \frac{V}{n.Q} = \frac{3000 \text{ m}^3}{3 \times 600 \text{ m}^3/\text{d}} = 1,67 \text{ d}$$

b) Concentration in each cell

The concentration in each cell is given by (see Table 2.5):

$$C = C_o/(1 + K.t_1)^n$$

For each of the 3 cells:

Formula	C (g/m^3)
$200/(1 + 0.40 \times 1.67)^1$	120
$200/(1 + 0.40 \times 1.67)^2$	72
$200/(1 + 0.40 \times 1.67)^3$	43

The final effluent concentration is equal to 43 g/m^3. As expected, this value is higher than that obtained in the most efficient system, represented by the plug flow (C_e = 27 g/m^3; Example 2.1), but is lower than that from the less efficient system, represented by a single complete-mix reactor (C_e = 67 g/m^3; Example 2.2).

The concentration profile along the three reactors is plotted below.

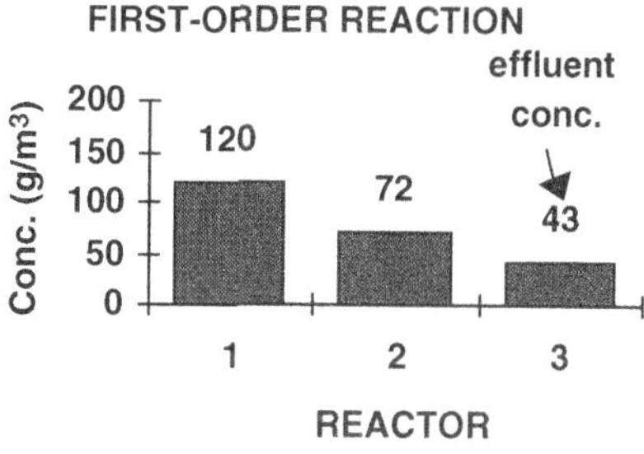

Example 2.3 (Continued)

c) Removal efficiency

The overall removal efficiency is:

$$E = (200 - 43)/200 = 0.785 = 78.5\%$$

The total hydraulic detention time is 5.0 days and the dimensionless product K.t is $0.4\,d^{-1} \times 5.0\,d = 2.0$. The efficiency of 78.5% can be also obtained from Figure 2.11, for K.t = 2.0 and n = 3 cells.

2.4.5 Dispersed flow

In real terms, the reactors that exist in practice do not behave exactly like the two idealised hydraulic models of plug flow and complete mix. However, these two ideal models configure an envelope, inside which the reactors can be found in practice. The reasons that cause real reactors not to follow the ideal models can be (Tchobanoglous & Schroeder, 1985):

- *Dispersion.* The dispersion is the longitudinal transportation of the material due to turbulence and molecular diffusion.
- *Hydraulic short circuits.* These take account of part of the flow and are the result of stratification, for instance due to a density difference, and not due to a physical characteristic of the system. The main effect is the reduction in the effective residence time.
- *Dead volumes.* The effect is similar to the short circuits (reduction of the effective residence time), but the causes are a function of the physical characteristics of the system. They occur in corners of tanks, underneath weirs and in the internal side of curves.

Consequently, the dispersed or arbitrary flow is a *non-idealised* case, and can be used in practice to describe flow conditions in most reactors. An approximation of the dispersed flow is represented by the system of cells in series, described in Item 2.4.4.

The mixing conditions in dispersed-flow reactors are characterised by a **Dispersion number**, defined as:

$$d = D/U.L \tag{2.28}$$

where:
   d = dispersion number (–)
   D = axial or longitudinal dispersion coefficient (L^2T^{-1})
   U = mean horizontal velocity (LT^{-1})
   L = reactor length (L)

Table 2.6. Typical values of d (= D/UL) for different
treatment units

Treatment unit	Range of d (= D/UL)
Rectangular sedimentation tanks	0.2–2.0
Aeration tanks for activated sludge	
– plug-flow type	0.1–1.0
– complete mix type	3.0–4.0 or more
– oxidation ditches	3.0–4.0 or more
Aerated lagoons	
– long, rectangular	0.2–1.0
– square format	3.0–4.0 or more
Non-aerated stabilisation ponds	
– long, rectangular	0.1–0.3
– square format	0.8–1.2

Source: Arceivala (1981)

In the two idealised reactors, there are the following limit conditions:

- *Plug flow*: *no dispersion* ($D = 0$ and $d = 0$)
- *Complete mixing*: *infinite dispersion* ($D = \infty$ and $d = \infty$)

The reactors found in practice have values of d situated between 0 and ∞. The value of d can be estimated by the use of tracers, a topic that is outside the scope of this text. The references (Grady & Lim, 1980; Arceivala, 1981; von Sperling, 1983b; Tchobanoglous & Schroeder, 1985; Viessman & Hammer, 1985) present the methodology and examples for this application. Table 2.6 presents ranges of d values for various treatment units.

Treatment units that have d values around *0.2 or less* are closer to *plug flow*. Conversely, units with values of d around *3.0 or more* can be considered to approach *complete mix*. Among the factors that can affect the dispersion of the treatment units, the following can be listed (Arceivala, 1981):

- Scale of the mixing phenomenon;
- Geometry of the unit;
- Energy introduced per unit volume (mechanical or pneumatic);
- Type and arrangement of the inlets and outlets;
- Inflow velocity and its fluctuations;
- Density and temperature differences between inflow and reactor contents
- Reynolds number (which is a function of some of the factors listed above).

It is important to note that the characterisation between plug-flow and complete-mix conditions is also a function of the dynamics of the constituent being analysed. For example, oxidation ditches behave like complete-mix reactors for most of the variables, such as suspended solids and BOD. Samples collected along its length will give approximately the same concentrations. However, for constituents that exhibit fast dynamics, the situation is different. Dissolved oxygen (DO) in activated sludge reactors presents very rapid dynamics, with fast increases or decreases in its concentration. For this reason, DO concentrations are high in the vicinity of the aerators, decreasing due to the bacterial consumption as the liquid flows along the ditch, until it reaches the next aerator. Therefore, there is a gradient of the longitudinal DO concentration along the tank, what characterises a regime approaching plug flow.

The analytical solution of the equation for dispersed flow with first-order kinetics was proposed by Wehner and Wilhem in 1956. For other reactions different from first order, numerical solutions are necessary. The equation for first-order reactions is:

$$C = C_0 \cdot \frac{4ae^{1/2d}}{(1 + a)^2\ e^{a/2d} - (1 - a)^2\ e^{-a/2d}}$$

$$a = \sqrt{1 + 4K.t.d}$$

(2.29)

where:
- d = dispersion number = D/UL = $D.t/L^2$ (–)
- D = coefficient of longitudinal dispersion (m^2/d)
- U = average flow velocity in the reactor (m/d)
- L = travel distance (m)
- t_h = hydraulic detention time (= V/Q) (d)
- K = removal coefficient (d^{-1})
- C = effluent concentration (g/m^3)
- C_0 = influent concentration (g/m^3)

The advantage of this equation is that it allows a continuous solution between the limits of plug flow and complete mix. When d is small, Equation 2.29 gives results very close to the specific equation for plug flow. On the other hand, when d is very high, Equation 2.29 produces similar values to those obtained from the equation for complete mix.

The use of the Wehner–Wilhem equation can be facilitated through the employment of graphs. Figure 2.12 presents a graph of the dimensionless product $K.t_h$ versus the removal efficiency $[(C_0 - C)/C_0]$, following first-order kinetics. Various curves are presented, all situated inside the envelope represented by d varying from 0 (plug flow) to ∞ (complete mix). In the design of a treatment unit, given the values of d and K and for a desired removal efficiency, the necessary hydraulic detention time t_h (and as a result the reactor volume) can be obtained. Similarly, if it is desired to estimate the efficiency of a reactor with a pre-defined volume, knowing K, t_h and d, the efficiency can be readily obtained from the graph.

Figure 2.13 presents the same family of curves, for a broader scale of $K.t_h$, and for a greater efficiency range (applicable, for instance, to coliform removal,

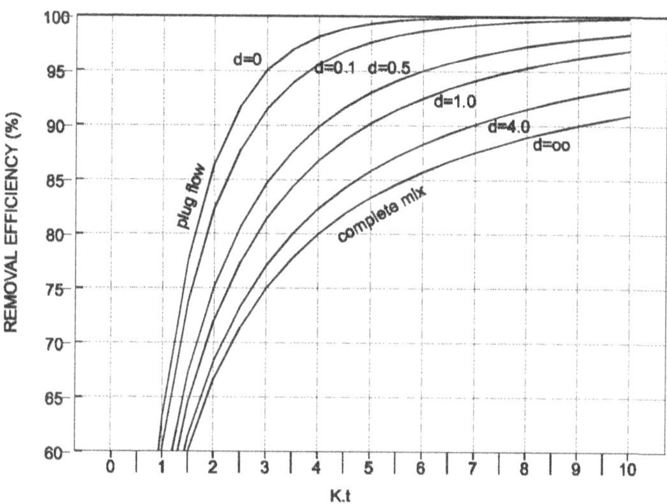

Figure 2.12. Removal efficiency (first order reaction) according to the Wehner–Wilhem equation for dispersed flow

in which high efficiencies are required). The removal efficiencies are presented in a logarithmic scale and also in terms of log units removed. An efficiency of $E = 90\%$ corresponds to a removal of 1 logarithmic unit; $E = 99\% \rightarrow 2$ log units; $E = 99.9\% \rightarrow 3$ log units; $E = 99.99\% \rightarrow 4$ log units; $E = 99.999\% \rightarrow 5$ log units, or:

$$\boxed{\text{Log units removed} = -\log_{10}[(100 - E)/100]} \qquad (2.30)$$

where:

E = removal efficiency, expressed in percentage (%)

The interpretation of Figures 2.12 and 2.13 for constituents that decay following first-order kinetics leads to the following points (Arceivala, 1981):

- For a given value of $K.t_h$, the reactors that approach plug flow always give higher efficiencies than the reactors that approach complete mix.
- A complete-mix reactor or even a relatively well-mixed reactor ($d > 4.0$) is incapable of giving a removal efficiency greater than 97% for values of $K.t_h$ less than 20.
- Very high efficiencies (greater than 99%), can only be reached if the system approaches plug-flow conditions (if the removal coefficient K is not especially high, or if the adoption of very high detention times is not desired).

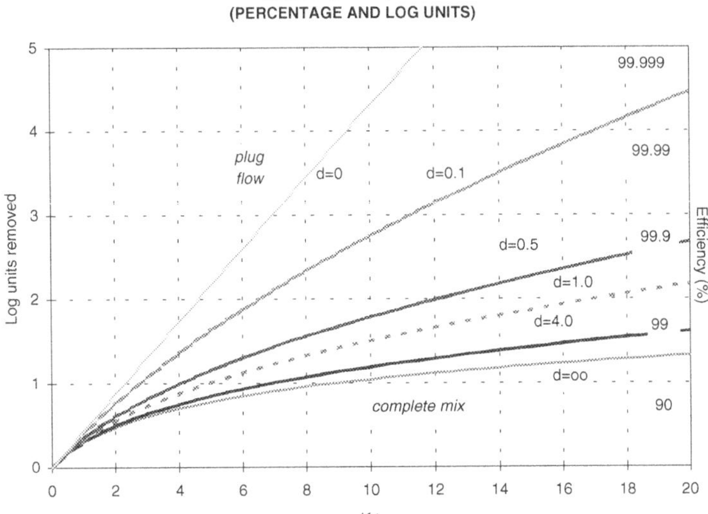

Figure 2.13.   Removal efficiency following a first-order reaction, in a dispersed-flow reactor, for different values of K.t$_h$ and d

---

Example 2.4

A reactor has the same volume as the reactor of Example 2.1 (3,000 m^3). The influent also has the same characteristics as the referred example (flow = 600 m^3/d; influent substrate concentration = 200 g/m^3).

Calculate the effluent concentration from the reactor. Assume that the dispersion number is 1.0 and that the substance is biodegradable with first order removal (K = 0.40 d^{-1}).

Solution:

a)   Hydraulic detention time

The detention time is calculated in the same way as in Example 2.1, that is, t$_h$ = 5 days.

b)   Calculation of the parameter *a*

According to Equation 2.29:

$$a = (1 + 4.K.t_h.d)^{0.5} = (1 + 4 \times 0.4 \times 5 \times 1.0)^{0.5} = 3.0$$

Example 2.4 (Continued)

c) Calculation of the effluent concentration

According to Equation 2.29:

$$C = C_0 . \frac{4ae^{1/2d}}{(1+a)^2 e^{a/2d} - (1-a)^2 e^{-a/2d}}$$

$$= 200 . \frac{4 \times 3.0 . e^{1/(2 \times 1.0)}}{(1+3.0)^2 . e^{3.0/(2 \times 1.0)} - (1-3.0)^2 . e^{-3./(2 \times 1.0)}}$$

$$= 200 \times 0.28 = 56 \text{ g/m}^3$$

This value is between the values obtained for a plug-flow reactor ($C_e = 27\text{g/m}^3$; Example 2.1) and complete-mix reactor ($C_e = 67\text{g/m}^3$; Example 2.2), although it is closer to a complete-mix reactor (because of the relatively high dispersion number).

The same value can be obtained from Figure 2.12. For K.$t_h = 0.4 \times 5 = 2.0$ and d $= 1.0$, a removal efficiency of 72% is obtained. With a removal of 72%, the remaining concentration is 28%, what corresponds to the value of 0.28 obtained in the second term on the right-hand-side of the Wehner–Wilhem equation above. Therefore, $C_e = 200 \times 0.28 = 56\text{g/m}^3$.

2.4.6 Cells in parallel

A treatment system is frequently composed of cells in parallel. Figure 2.14 shows a possible arrangement of cells in parallel.

With cells in parallel, the following points should be noted (Arceivala, 1981):

*   The cells can be equal or different in size, since they operate independently.
*   Even if the cells are of different sizes, they can be operated with the same detention time through the individual adjustment of each inlet flow.

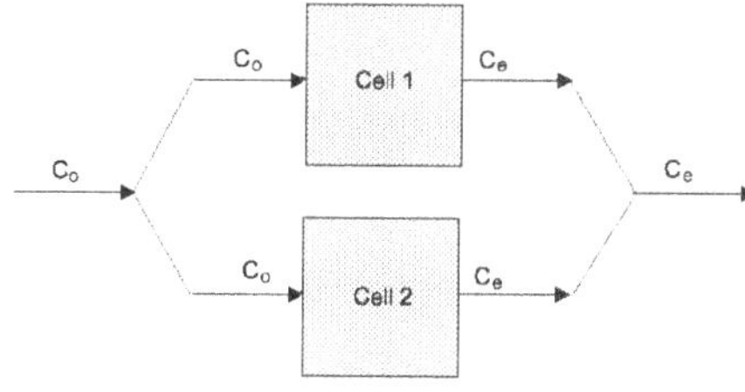

Figure 2.14.   Schematic arrangement of cells in parallel

- Each cell can be designed individually using the dispersed-flow model and adequate values of d for each cell. The d values can vary from cell to cell.
- If each cell behaves as a complete-mix reactor, the final effluent will have the same concentration of that from a reactor with only one cell and a volume equal to the total volume of all cells. Therefore, the subdivision of a reactor (complete mix) into parallel cells (each one complete mix) does not affect the effluent concentration.
- However, the point above will not apply for reactors modelled according to dispersed flow, since the subdivision of a reactor into smaller reactors in parallel can lead to reactors with different geometries from the original large reactor. Therefore, the smaller reactors will have different dispersion numbers and, as a result, different effluent concentrations.
- For a given total volume, the substrate removal efficiency in first-order reactions is lower for cells in parallel than for cells in series. However, a parallel arrangement is frequently convenient due to reasons such as operational flexibility, operation continuity even with the closing of one unit, sludge removal etc.

2.4.7 Cells in series with incremental feeding

When there is an arrangement of cells *in series*, it is possible that the inflow distribution is split between the various cells. Hence, each cell is fed not only by the effluent from the upstream cell, but also by a fraction of the general influent. In activated sludge the denomination *step feed* has been employed to classify plug-flow reactors or cells in series that receive this type of incremental feed. Such an arrangement can also be used in stabilisation ponds and trickling filters. Figure 2.15 shows possible arrangements of cells in series with step feed.

In an arrangement in series with n cells of equal or different volumes with incremental feeding, all the liquid fractions do not receive the same treatment exposure. The first fraction receives treatment in all the cells; the second fraction is treated in n − 1 cells; the third in n − 2 cells and so on. When the cells have the same volume and receive the same fraction of the total flow, and decay follows a

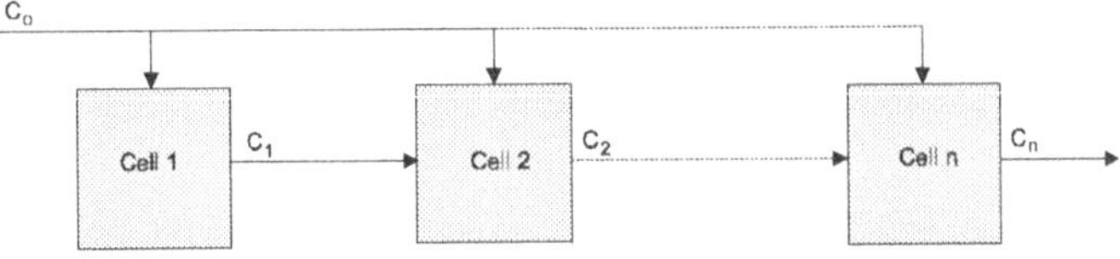

Figure 2.15.   Schematic arrangement of cells in series with step feed

first-order reaction, the effluent concentration is given by (Arceivala, 1981):

$$C_e = \frac{C_o}{1 + K.\left(\frac{n.V_1}{Q}\right)}$$

(2.31)

where:

    n = number of cells in series with incremental feeding (–)
    V_1 = volume of each cell (assuming equal volumes) (m³)
    Q = total influent flow in the system (m³/d)

It can be seen that such an arrangement has the same efficiency of a single complete-mix cell with an equivalent total volume. In other words, a reactor with incremental feeding behaves like a complete-mix reactor. In terms of efficiency, the incremental feeding loses the benefits from the arrangement in series. Obviously, other reasons of practical and operational order may justify the inclusion of this option, principally a greater operational flexibility. In the cases when the inlet end of a reactor or the first cell are overloaded, such flexibility can contribute to the control of this localised overload.

If the cells have different volumes and flows, the calculation can be done individually in each cell through the individual mass balances. If convenient, the dispersed flow model can be adopted for each cell with the corresponding d value.

2.4.8 Cells in series and in parallel

With the aim of having greater operational flexibility, the arrangement of cells in series and in parallel is frequently used. Thus, there are the benefits of efficiency with the arrangement in series and of flexibility with the arrangement in parallel. Figure 2.16 shows a typical series/parallel arrangement.

Since the units in parallel do not interfere in the efficiency, the calculations of the effluent concentration can be done using the formulas for cells in series (Table 2.5), adopting the complete-mix model and the corresponding value of

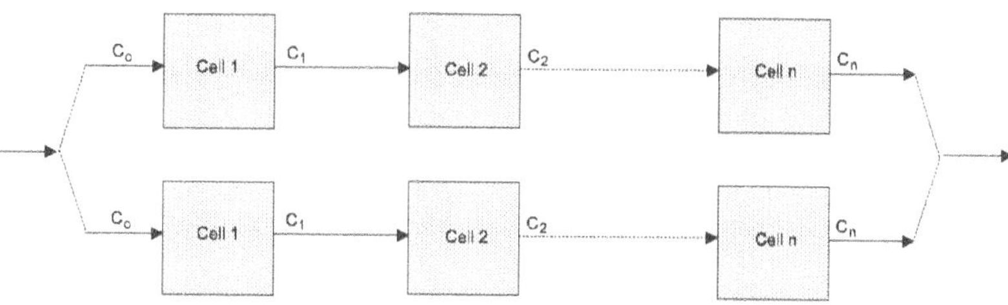

Figure 2.16. Schematic arrangement of cells in series and in parallel

n (number of cells in series in each line). Alternatively, the dispersed-flow model can be used with the appropriate value of d for each cell.

2.4.9 The influence of recirculation

In plug-flow or cells-in-series systems, a fraction of the effluent can be recirculated to the inlet of the reactor. Depending on the application, this fractions can be lower, equal or greater than the influent flow.

The recirculation is an inherent component of various treatment processes, such as activated sludge and high rate trickling filters. In the case of the activated sludge system, the recirculation is specific, since the recycled liquid has different characteristics (sludge removed from the bottom of the secondary settling tank). Therefore, the characteristics of the return (recycle) sludge are different from the effluent from the reactor, principally in terms of the concentration of the suspended solids. Given the importance of this phenomenon and the complexity of its interactions with the reactor, the sludge recirculation in activated sludge systems is not covered in the present section.

In any situation, when making a mass balance on the reactor, the following points must be taken into consideration:

- The incoming flow to the reactor is equal to the sum of the influent flow (Q_o) and the recycled flow (Q_r).
- The concentration in the incoming flow to the reactor is given by a weighted average between the influent and the recycled flows, according to:

$$\boxed{C_0' = \frac{Q_0.C_0 + Q_r.C_r}{Q_0 + Q_r}} \qquad (2.32)$$

where:
  C_0' = incoming concentration in the reactor (mixture of the influent and the recycled flows) (g/m^3)
  C_0 = concentration in the influent to the system (g/m^3)
  C_r = concentration of the recycled liquid (g/m^3)
  Q_0 = influent flow to the system (m^3/d)
  Q_r = recycled flow (m^3/d)

- The outgoing flow from the reactor is equal to the sum of the influent flow (Q_o) and the recycled flow (Q_r).
- The concentration in the reactor outlet is obtained using the pertinent equations, as a function of the reaction order and the hydraulic model adopted.

In the case of activated sludge, in which the sludge comes from the bottom of the secondary settling tank, the concentrations of the suspended solids in the return sludge are much higher than in the influent concentration. In these conditions, the weighted average leads to an incoming value C_0' greater than in the

Table 2.7. Required hydraulic detention time to obtain an effluent concentration C_e (steady state)

Reaction order	Hydraulic detention time	
	Complete mix	Plug flow
0	$(C_e - C_o)/K$	$(C_e - C_o)/K$
1	$[(C_o/C_e) - 1]/K$	$[\ln(C_o/C_e)]/K$
Saturation	$(C_o - C_e).(K_s + C_e)/(K.C_e)$	$[(K_s.\ln(C_o/C_e) + C_o - C_e]/K$

Source: Benefield and Randall (1980), Tchobanoglous and Schroeder (1985)
K = reaction coefficient
K_s = half-saturation coefficient

influent. Conversely, the substrate concentration is lower than in the influent, and the recirculation contributes to the reduction of the influent concentration.

In a generic system, in which the recirculation is taken directly from the effluent from the reactor (with low substrate concentration), the recirculation becomes responsible for a dilution of the influent substrate concentration. Consequently, the efficiency of the system is reduced, in the case of first or higher order reactions.

2.4.10 Comparison between the reactor types

The comparison between the performances of the various types of reactor is an important topic in the analysis and design of a wastewater treatment plant. As previously mentioned, the following generalisations can be made, assuming steady-state conditions.

- *Conservative substances*: plug-flow reactors, cells in series and complete-mix reactors present the same performance.
- *Biodegradable substances with a zero-order reaction*: plug-flow reactors, cells in series and complete-mix reactors present the same performance.
- *Biodegradable substances with a first-order reaction*: the plug-flow reactor presents the highest efficiency, followed by the cells-in-series system. The single complete-mix reactor is the least efficient.

The statement that a system is more efficient than another implies that, if both present the same effluent concentration, the less efficient system requires a higher detention time. In other words, the less efficient system must have a larger reactor volume. This consideration is of great importance in the design of a treatment plant. Table 2.7 presents a summary of the equations used to calculate the detention time required to obtain a certain concentration C_e in the effluent.

The interpretation of Table 2.7 leads to the following points:

- *Zero-order reactions*. For zero-order reactions, the required hydraulic detention is the same.
- *First-order reactions*. For first-order reactions, the application of the pertinent formulas leads to the requirement of the greatest detention times for the complete-mix system.

Table 2.8. Relative volumes (expressed as $K.t_h$) required for various removal efficiencies. First-order reactions (steady state)

Number of reactors in series	Relative volume (dimensionless product $K.t_h$)					
	85% efficiency	90% efficiency	95% efficiency	99% efficiency	99.9% efficiency	99.99% efficiency
1	5.7	9.0	19	99	999	9999
2	3.2	4.3	7.0	18	61	198
4	2.5	3.1	4.5	8.6	18	36
∞ (plug flow)	1.9	2.3	3.0	4.6	6.9	9.2

Source: Arceivala (1981), Metcalf & Eddy (1991)

- *Saturation reactions.* For saturation reactions, the result depends on the relative value of K_s with respect to C_e. When $C_e >> K_s$, the reaction tends to zero order and the reactor volumes for complete mix and plug flow are approximately the same. When $C_e << K_s$, the reaction tends to first order and the complete-mix system requires larger volumes than the plug-flow system.

Table 2.8 presents the relative volumes necessary for obtaining different removal efficiency values (assuming ideal complete-mix reactors and a first-order reaction). The table gives values of the dimensionless product $K.t_h$. Based on the desired efficiency and on the product $K.t_h$ (after knowing or estimating K), the required detention time can be obtained. With the detention time and the flow, the volume can be determined ($V = t_h.Q$).

It is confirmed in this table the fact that, for first-order reactions, the plug flow requires the lowest volume for a given efficiency. The greater the required efficiency, the higher the ratio (complete mix volume) / (plug flow volume). For an efficiency of 85% this ratio is 3.0 ($= 5.7/1.9$), that is, the volume required by a single cell is three times higher than that of the plug flow. However, for an efficiency of 99% this ratio becomes 21.5 ($= 99/4.6$). The simple subdivision of the total volume in 2 cells changes these ratios to 1.7 ($= 3.2/1.9$) and 3.9 ($= 18/4.6$), respectively. However, it should be remembered that these considerations are based on the assumption of ideal complete-mix and plug-flow reactors, which is hardly achieved in practice.

2.4.11 Comparison between first-order reaction coefficients in different hydraulic models

2.4.11.1 Estimation of the reaction coefficients in existing reactors

Table 2.9 presents the formulae for the estimation of the effluent concentration of a first-order decay pollutant, as a function of the hydraulic regime assumed for the reactor. For an existing reactor, the coefficient K can be calculated by rearranging the equations in Tables 2.3 and 2.4, and making K explicit, provided the influent concentration C_0, the effluent concentration C and the detention

Table 2.9. Formulas for the estimation of the first-order decay coefficient K, for different hydraulic regimes

Hydraulic regime	Formula for the decay coefficient (K)	
	Equation	Equation number
Plug flow	$K = \dfrac{-\ln{(C/C_0)}}{t}$	(8.33)
Complete mix	$K = \dfrac{(C_0/C) - 1}{t}$	(8.34)
Dispersed flow	K value not explicit. Solve by iteration (trial-and-error or error function minimisation)	–

time t (also the dispersion number d, for dispersed-flow models) are known or have been determined. The analysis undertaken in the present section is based on von Sperling (2002).

For a given removal efficiency, the estimation of K based on the detention time and on the influent and effluent concentrations on an existing reactor leads to the two following divergent situations:

- adoption of the **complete-mix** (CSTR) model leads to K values which are **greater** than those found for dispersed flow
- adoption of the **plug-flow** model leads to K values which are **lower** than those found for dispersed flow

The following example will help to clarify the point. An existing reactor has the following average values of performance indicators: (a) influent coliform concentration: $C_0 = 1 \times 10^7$ FC/100 ml; (b) effluent coliform concentration: $C = 2.13 \times 10^5$ FC/100 ml; (c) detention time: t = 30 days; (d) dispersion number: d = 0.5. Use of Equations 2.33 and 2.34 will lead to the K coefficients for plug flow and complete mix, respectively. An iterative process of trial-and-error will lead to the K coefficient for dispersed flow. The following K values are obtained: (a) plug flow: $K = 0.13$ d^{-1}; (b) CSTR: $K = 1.53$ d^{-1}; (c) dispersed flow: $K = 0.30$ d^{-1}. As can be seen, for the same reactor and the same kinetics, different K values are obtained in practice, depending on the hydraulic regime assumed.

In principle, there should be only one coefficient, representing the decay of the constituent, according to its kinetics. However, the inadequacy of idealised models in representing in a perfect manner the real hydraulic pattern in the reactor leads to the deviations that occur in practice. The reason for the differences observed in the example above is that, since complete-mix reactors are the least efficient for first-order removal kinetics, the lower efficiency is compensated by a higher K value. Conversely, since plug-flow reactors are the most efficient reactors, the K value is reduced to produce the same effluent quality. Depending on the length/breadth (L/B) ratio of the reactor (dispersion characteristics), the deviation can be very large, inducing considerable errors in the estimation. Naturally,

K coefficients for dispersed flow are assumed to best represent reality and the true reaction kinetics. However, the confidence on K values for dispersed flow relies very much on the confidence on the assumed or determined values of the dispersion number d.

These divergences have been the subject of considerable confusion in the literature, when expressing K values. Reported K values usually show considerable variations, a large part of which can be attributed to inadequate consideration of the hydraulic regime of the reactor.

2.4.11.2 Relationship between K for idealised regimens (complete mix and plug flow) and K for dispersed flow

The present section, also based on von Sperling (2002), describes the relationship between first-order K values for the idealised flow patterns (CSTR $-$ K_{CSTR} and plug flow $-$ K_{plug}), and K for the general flow pattern, dispersed flow (K_{disp}).

The following explanation demonstrates the methodology applied for the CSTR regime. A similar methodology, using the appropriate equations, was also used for the plug-flow regime. Using the relevant equations for estimating the effluent concentrations under complete mix and dispersed flow, it was calculated, for different values of the dimensionless product $K_{disp}.t$ and dispersion number d, the corresponding K_{CSTR}, which yields the same efficiency of removal (first-order kinetics). The dispersion numbers d ranged from extremely high values (100,000, representing complete-mix conditions) to extremely low values (0.001, representing plug-flow conditions).

The results are presented in Table 2.10, showing the ratio between the K for CSTR and K for dispersed flow (K_{CSTR}/K_{disp}). The interpretation of the table is as follows. The same reactor analysed in the previous section, with d $= 0.5$, detention time t $= 30$ days and $K_{disp} = 0.3$ d^{-1} has the dimensionless product $K_{disp}.t = 0.3 \times 30 = 9.0$. For d $= 0.5$ and $K_{disp}.t = 9$, the table shows that the K_{CSTR} is equal to 5.144 times K_{disp}. In other words, K_{CSTR} is $5.144 \times 0.3 = 1.54$ d^{-1}. This value is, apart from rounding values, the same obtained in the previous section (1.53 d^{-1}), indicating the applicability of the table. The estimation of the removal efficiency using the dispersed-flow model (using K_b for dispersed flow) and the CSTR model (using K_b for CSTR) will lead to the same results.

Table 2.11 shows the corresponding values for the plug-flow model. In the same example, it is seen from Table 2.11 that, for d $= 0.5$ and $K_{disp}.t = 9$, K_{plug} is 0.430 times K_{disp}. Therefore, $K_{plug} = 0.430 \times 0.3 = 0.13$ d^{-1} (which is exactly the same value determined in the previous section).

Figure 2.17 illustrates the data from Tables 2.10 and 2.11. It can be clearly seen that, for the CSTR regime, the smaller the dispersion number d, the greater the departure between K_{CSTR} and K_{disp}. Conversely, for the plug-flow regime, the greater the Dispersion number d, the greater is the departure between K_{plug} and K_{disp}. The departure also increases with the detention time t. Another point to be observed is that the relative departures can be much larger for the CSTR regime

Table 2.10. Ratio between the K coefficients obtained for the complete-mix model and the dispersed-flow model, for different values of the dispersion number d and of the product $K_{disp}.t$

| | Ratio K_{CSTR}/K_{disp} | | | | | | | |
$K_{disp}.t$	d = 100,000	d = 4	d = 1	d = 0.5	d = 0.2	d = 0.1	d = 0.02	d = 0.001
0	1.000	1.000	1.000	1.000	1.000	1.000	1.000	1.000
1	1.000	1.040	1.140	1.230	1.400	1.520	1.670	1.715
2	1.000	1.075	1.290	1.515	1.950	2.320	2.940	3.180
3	1.000	1.120	1.457	1.833	2.677	3.550	5.380	6.300
4	1.000	1.163	1.635	2.213	3.658	5.393	10.150	13.175
5	1.000	1.210	1.832	2.646	4.950	8.180	19.440	28.800
6	1.000	1.255	2.043	3.150	6.617	12.283	37.620	64.667
7	1.000	1.300	2.271	3.729	8.814	18.214	73.000	149.000
8	1.000	1.346	2.525	4.388	11.600	26.813	141.000	350.000
9	1.000	1.394	2.789	5.144	15.156	39.111	272.780	831.111
10	1.000	1.444	3.080	6.010	19.660	56.500	524.000	1995.000

Table 2.11. Ratio between the K coefficients obtained for the plug-flow model and the dispersed-flow model, for different values of the dispersion number d and of the product $K_{disp}.t$

| | Ratio K_{plug}/K_{disp} | | | | | | | |
$K_{disp}.t$	d = 100,000	d = 4	d = 1	d = 0.5	d = 0.2	d = 0.1	d = 0.02	d = 0.001
0	1.000	1.000	1.000	1.000	1.000	1.000	1.000	1.000
1	0.695	0.715	0.762	0.805	0.878	0.926	0.984	1.000
2	0.551	0.578	0.640	0.699	0.797	0.868	0.967	1.000
3	0.463	0.493	0.562	0.626	0.736	0.820	0.950	1.000
4	0.404	0.435	0.506	0.574	0.689	0.782	0.935	0.999
5	0.359	0.392	0.465	0.532	0.652	0.749	0.920	0.998
6	0.325	0.358	0.432	0.500	0.620	0.721	0.907	0.997
7	0.298	0.331	0.405	0.473	0.593	0.696	0.894	0.996
8	0.276	0.309	0.383	0.450	0.569	0.674	0.882	0.995
9	0.257	0.291	0.364	0.430	0.549	0.654	0.870	0.994
10	0.241	0.274	0.347	0.413	0.530	0.636	0.859	0.993

than for the plug-flow regime, indicating that an even greater caution needs to be exercised when applying the CSTR model.

In reactors without mechanical mixing, the lowering of the dispersion number d occurs with the increase in the length/breadth (L/B) ratio. In other words, a baffled reactor is likely to have a low value of d. Under these circumstances, utilisation of the CSTR model will be completely inadequate, due to the large difference between K_{CSTR} and $K_{dispersed}$, the latter being naturally expected to be a better predictor of the actual behaviour in the reactor. In this baffled reactor, use of the CSTR model for design purposes, adopting 'typical' values of K_{CSTR} from the literature will lead to an underestimation of the removal efficiency in the reactor. On the other hand, for an existing baffled reactor, the calculation of the coefficient

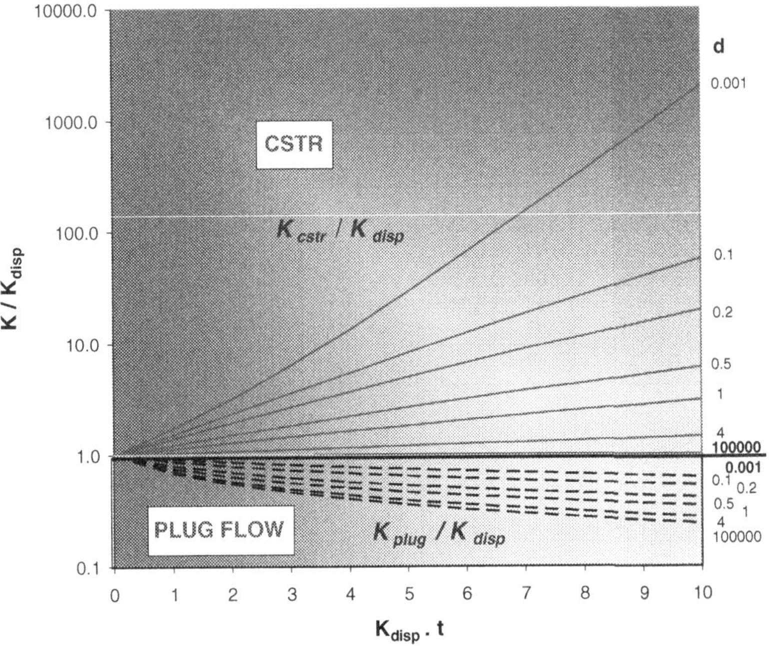

Figure 2.17.   Relationship between coefficients K for CSTR and plug flow with the coefficient K for dispersed flow, as a function of the dispersion number d and the hydraulic detention time t.

K using the CSTR model will lead to an overestimation of the K coefficient, in order to compensate for the inherent lower efficiency associated with the CSTR model.

In order to extend the applicability of Tables 2.10 and 2.11, a regression analysis was done by von Sperling (2002), having as dependent variable the ratio between the K value for the idealised regime (CSTR or plug flow) and the K value for the general regime (dispersed flow). The dependent variables were then K_{CSTR} / K_{disp} and K_{plug} / K_{disp}. The independent variables were the dimensionless product $K_{disp}.t$ and the dispersion number d. Two regression analyses were done, each one having different applicability ranges. The equations of best fit obtained were:

Wider applicability range (d from 0.1 to 4.0; $K_{disp}.t$ from 0 to 10; n = 55 values from Tables 2.10 and 2.11):

* *For CSTR* (R^2 = 0.994):

$$\frac{K_{CSTR}}{K_{disp}} = 1.0 + \left[0.0020 \times (K_{disp}.t)^{3.0137} \times d^{-1.4145}\right] \qquad (2.35)$$

- *For plug flow* ($R^2 = 0.956$)

$$\frac{K_{plug}}{K_{disp}} = 1.0 - \left[0.2414 \times (K_{disp}.t)^{0.4157} \times d^{0.1880}\right] \qquad (2.36)$$

Narrower applicability range (d from 0.1 to 1.0; $K_{disp}.t$ from 0 to 5; n = 24 values from Tables 2.10 and 2.11):

- *For CSTR* ($R^2 = 0.994$)

$$\frac{K_{CSTR}}{K_{disp}} = 1.0 + \left[0.0540 \times (K_{disp}.t)^{1.8166} \times d^{-0.8426}\right] \qquad (2.37)$$

- *For plug flow* ($R^2 = 0.987$)

$$\frac{K_{plug}}{K_{disp}} = 1.0 - \left[0.2425 \times (K_{disp}.t)^{0.3451} \times d^{0.3415}\right] \qquad (2.38)$$

All fits were very good, as indicated by the high R^2 values obtained. The reason for having equations for two applicability ranges is that the wider-range equation is not very accurate for lower values of d or $K_{disp}.t$, therefore making the narrower-range equations more adequate under these circumstances. From the equations, it is seen that K_{CSTR}/K_{disp} *will always be greater than 1.0*, whereas K_{plug}/K_{disp} *will always be lower than 1.0.*

2.4.12 The influence of variable loads

2.4.12.1 General concepts

The comparison between the efficiencies presented in Section 2.4.10 was based on the steady-state assumption, in which the influent characteristics remain constant. In a wastewater treatment plant this constancy rarely occurs. The variation of the flow and concentration along the day is responsible for the fact that, in reality, the system always operates in a *dynamic state*. Besides this, various other factors can contribute to a greater variability, such as stormwater flow (especially in combined systems) and industrial discharges. The latter can occur without any established periodical pattern and can be responsible for shock loads at the works. The shock loads can be of various natures, such as hydraulic, organic, toxic, of a non-biodegradable substance, thermal etc. A wastewater treatment plant must be apt to receive overloads that occur routinely or frequently, as well as a major part of the unpredicted overloads.

In the situations in which this variability component is substantial, the conception of the system must take this fact into consideration, the importance of which could even surpass the efficiency considerations discussed in Section 2.4.10. The effects of shock loads are best evaluated through the study of transients, using dynamic mathematical models of the system. These simulations can use the typical or

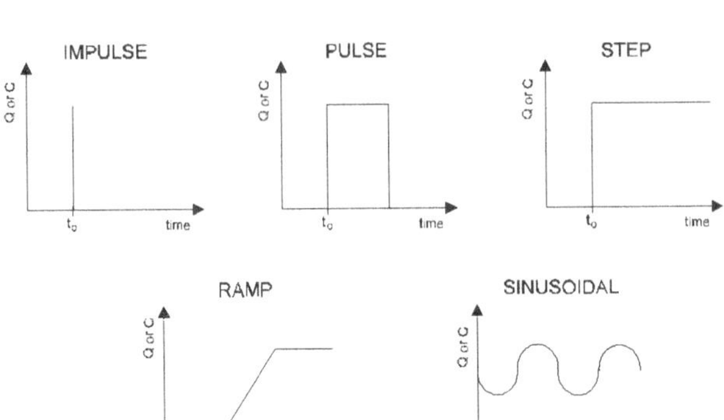

Figure 2.18. Transient analysis. Standardised influent variations.

expected variations of the influent characteristics, such as standardised variations. Some of the standardised variations of the influent characteristics normally used in transient analysis are (see Figure 2.18):

The analysis of these elements is outside the scope of this book (with the exception of the step function, covered in Sections 2.4.12.2 and 2.4.12.3). However, some generalisations can be made:

- *Toxic substances instantaneously added as spikes.* The peak in the effluent from the complete-mix reactor is the smallest, opposed to the plug-flow reactor, which presents the highest peak in the effluent. The good performance of the complete-mix reactor is caused by the large and instantaneous dilution provided at the entrance in the reactor. Additionally, the greater volume required for the single-cell complete-mix reactor contributes to the smoothing of the shock load. In the first reactor of a system with cells in series or at the head of a plug-flow reactor the toxic concentrations can be very high, due to the lower volumes involved.

- *Toxic substances with step increase.* The plug-flow reactor subjected to a step load of a conservative substance reaches a new equilibrium concentration after a time equal to 1 t_h. In the same period, the complete-mix reactor reaches only 63% of the equilibrium concentration, 3 t_h being required for the concentration in the reactor to reach 95% of the equilibrium concentration. This larger time can be fundamental for sustaining the system or for corrective operational control measures to be taken. Also in this case the larger volumes usually found in the complete-mix reactors contribute to a greater stability in the system.

- *Overload of biodegradable substances.* As already seen, for first-order reactions, the efficiency of a single complete-mix reactor is lower than that of a system in series or a plug flow. However, the consideration of the volume of the unit plays an important role in the case of transients. In the first cell of a system in series or in the head of a plug-flow reactor, due to the lower volume involved, the effect of the overload can be more deleterious. In aerobic systems, if the oxygenation capacity in these volumes is not sufficient, the organic overload could even lead to anaerobic conditions.

- *Hydraulic overload.* When a sudden increase in the flow occurs, a dilution of the reactor contents also takes place, which can be responsible for the washing-out of biomass from the reactor. With the decrease in biomass concentration, there is a resulting reduction in the efficiency of the system. The smaller the reactor volume, the greater is its susceptibility to this wash-out. For this reason, single-cell complete-mix reactors are more stable than systems in series or plug flow.

Depending on the way in which the final effluent quality is monitored, the concept of efficiency can vary:

- *Composite samples.* Systems that verify the final effluent quality through composite samples cannot detect the concentration peaks in the effluent. In these cases, the greater stability provided by the complete-mix system may not be apparent
- *Simple samples.* Systems that verify the final effluent quality through simple (instantaneous or grab) samples are subject to the collection of a sample at a moment of peak concentration in the effluent. This can be sufficient for a WWTP to be detected as infringing the discharge standards. In this situation, the stability provided by the complete-mix system will be apparent.

In summary, the selection between one type of reactor and another is a compromise between **mean efficiency** and **stability**. Each case must be analysed individually.

2.4.12.2 *Plug-flow reactor subjected to step variations in the influent concentration*

In an ideal plug-flow reactor, in the cases when the influent concentration increases instantaneously to a new level (at which it stays), the behaviour is very similar to that described for constant concentrations (steady state). The main difference is in the sense that the change takes place while the plug with the new concentration flows downstream. The particles downstream still have the old concentration, while the particles upstream are already with the new higher concentration. In this ideal plug-flow reactor, the effluent concentration will only be altered after the complete flow of the plug, which takes exactly the same time as the hydraulic detention time. Figure 2.19 illustrates the behaviour of a plug-flow reactor subjected to a step increase in the influent concentration of conservative constituents and constituents that decay according to 0 and 1st order kinetics.

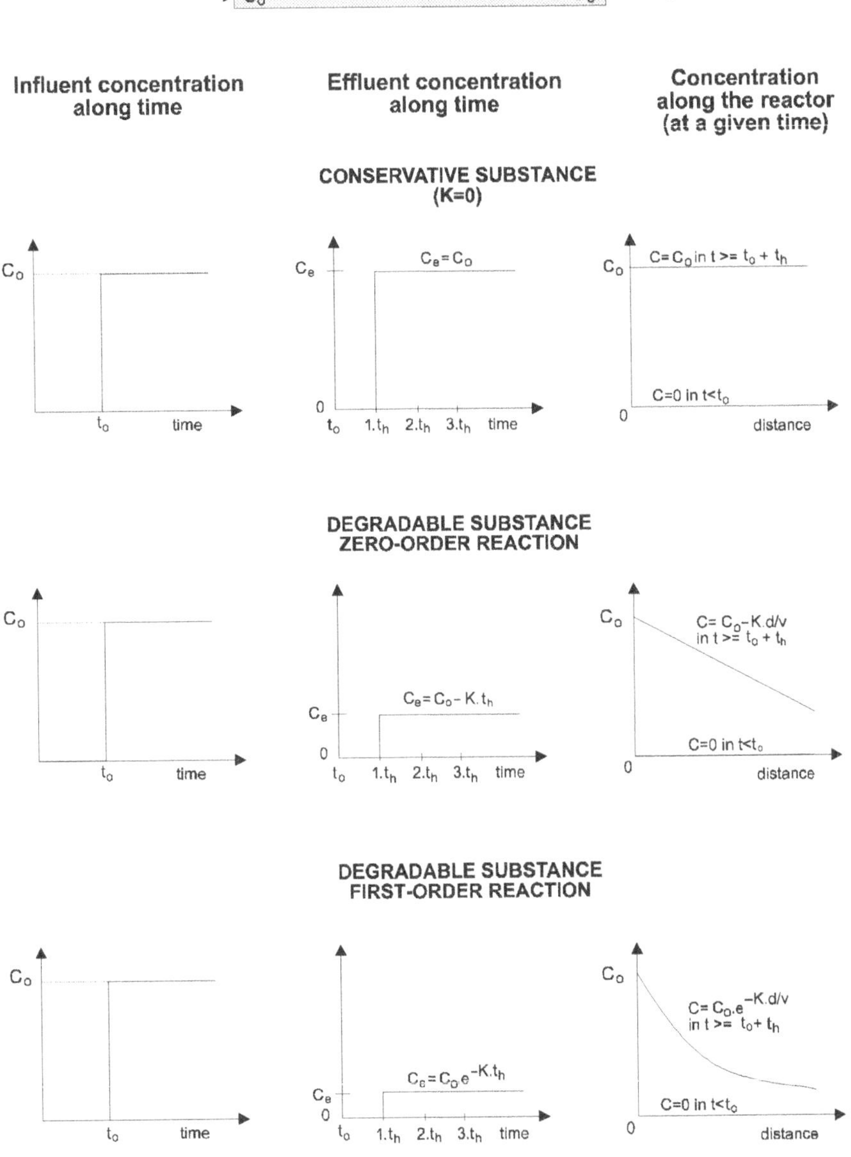

Figure 2.19. Transient analysis in a plug-flow reactor. Step increase in the concentration. (d = distance along the reactor).

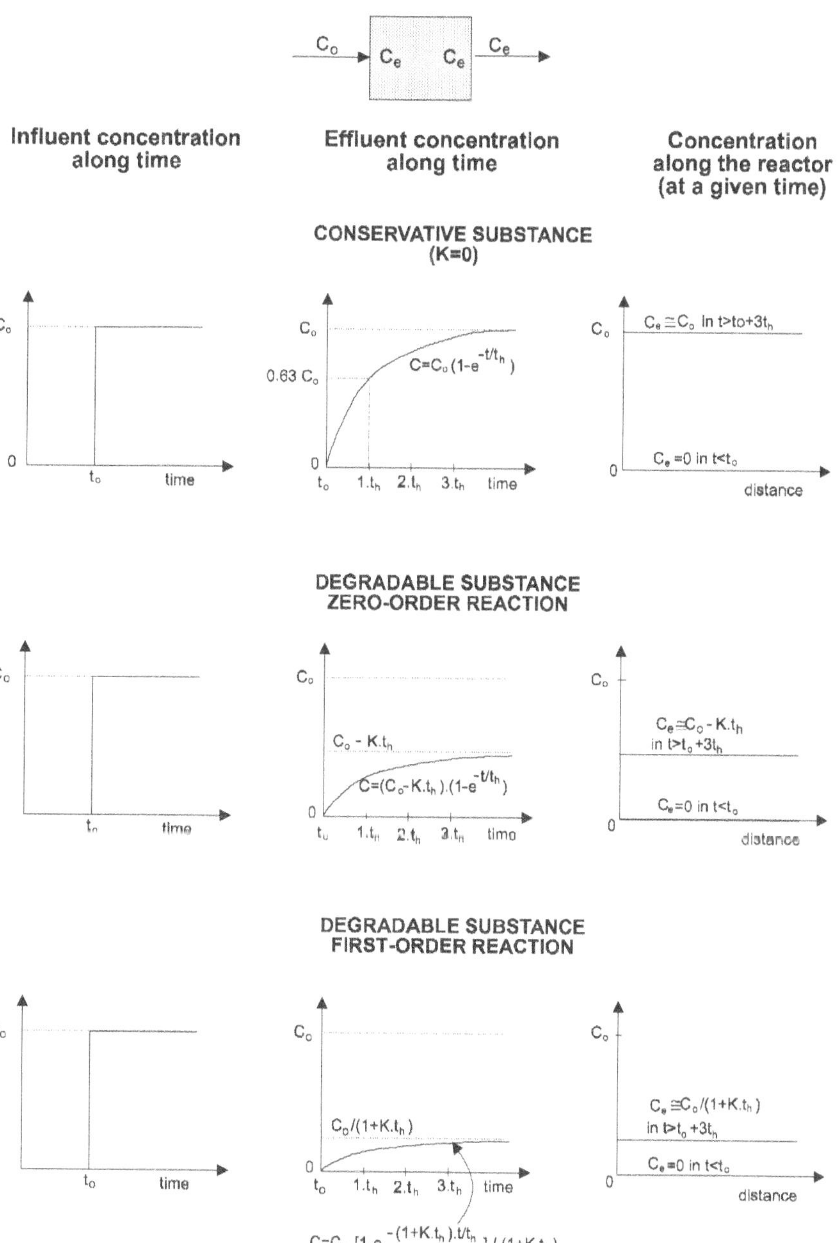

Figure 2.20.   Transient analysis in a complete-mix reactor. Step increase in the concentration. (d = distance along the reactor).

2.4.12.3 *Complete-mix reactor subjected to step variations in the influent concentration*

In the cases that the influent concentration increases instantaneously to a new level (at which it stays), the behaviour of the complete-mix reactor is essentially different from the plug-flow reactor for any reaction order. This is due to the hydraulic characteristics of the complete-mix reactor, in which the influent substance is immediately dispersed in the tank, instantaneously appearing in the effluent. With the continuous arrival of the new higher influent concentration, the effluent concentration continues to increase until the transient conditions stop and a new level is reached. The effluent concentration remains at this new level, because a new steady state has been reached, until there is a new change in the influent characteristics.

Figure 2.20 shows the behaviour of a complete-mix reactor subjected to a step increase in influent concentration of conservative constituents and constituents that decay according to 0 and 1st order kinetics. The equations presented for the transients are asymptotic with relation to the new equilibrium value. Thus, in strict mathematical terms, a new equilibrium value will never be reached. For conservative substances, the utilisation of the equation $C = C_0.(1 - e^{-t/th})$ leads to the following values of the ratio C/C_0 (remaining concentration / influent concentration).

t/t_h	1.0	2.0	3.0	4.0	5.0
C/C_0	0.63	0.86	0.95	0.98	0.99

It can be observed that, after a time corresponding to the hydraulic detention time, the concentration of a conservative substance is 63% of the new equilibrium concentration (which is equal to the new influent concentration in the case of conservative substances). After a time equal to three times the hydraulic detention time, the concentration is equal to 95% of the equilibrium concentration. Hence, in practical terms, it can be considered that after a period greater than $3t_h$, a new equilibrium concentration will be reached. In any of the three equations (conservative substances, zero and first order reactions) presented in Figure 2.20 for the step function, when the time tends to infinity, the equations are converted into the steady-state form (presented in Table 2.4).

3

Conversion processes of organic and inorganic matter

3.1 CHARACTERISATION OF SUBSTRATE AND SOLIDS

3.1.1 Introduction

In sewage treatment, there is an interaction between various mechanisms, some occurring simultaneously, and others sequentially. The microbial action starts in the sewerage system and reaches its maximum in the sewage treatment works. In treatment plants, the conversion of organic matter to more oxidised or reduced forms takes place. Under aerobic conditions there is the oxidation of the organic matter (*carbonaceous matter*), that is, the organic carbon is converted into its most oxidised form (CO_2: carbon in the oxidation state of 4+). Under anaerobic conditions, the conversion reaction of the organic matter leads to the most oxidised form of carbon (CO_2), but also to its most reduced form (CH_4: carbon with an oxidation state of 4−). In sewage treatment under aerobic conditions, the conversion of ammonia (*nitrogenous matter*) into more oxidised forms of nitrogen ($NO_3{}^-$) can take place, and, under anoxic conditions, the subsequent conversion of these to reduced forms (N_2) can also happen. Biological wastewater treatment therefore includes oxidation (increase of the oxidation state) and reduction (decrease in the oxidation state) reactions.

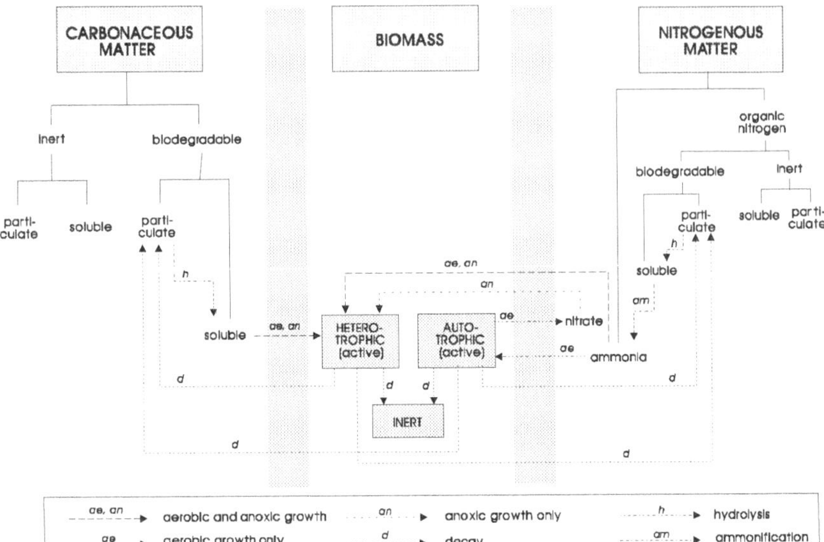

Figure 3.1.   Subdivisions and transformations of carbonaceous and nitrogenous matter

The main transformation phenomena, together with the participation of the biomass, are presented in a schematic form in Figure 3.1 (interpreted from the mathematical model of the IAWPRC, 1987). It can be observed that there is a high complexity in the interrelation of the various compounds and biomass. The following items describe the main concepts and mechanisms related with the interaction of the biomass (central area of the figure) with the *carbonaceous* and *nitrogenous* (side areas) matter.

3.1.2 Characterisation of the carbonaceous matter

The carbonaceous matter (based on organic carbon) present in the wastewater to be treated can be divided in terms of biodegradability into (a) *inert* or (b) *biodegradable*.

- The **inert organic matter** (non-biodegradable) passes through the treatment system without changing its form. Two fractions can be identified with respect to the physical state:
    - *Soluble*. The non-biodegradable soluble organic matter does not undergo transformations and leaves the system with the same concentration that it entered.
    - *Particulate*. The non-biodegradable particulate organic matter (suspended) is involved by the biomass and is removed together with the

sludge (excess sludge or the sludge that settles at the bottom of the reactors).

- The **biodegradable organic matter** is changed in its passage through the system. Two fractions can be identified, related to the biodegradability, which is also dependent on the physical state:
  - *Rapidly biodegradable*. This fraction is usually in a *soluble* form and consists of relatively simple molecules. These molecules can be directly used by the heterotrophic bacteria.
  - *Slowly biodegradable*. This fraction is usually in a *particulate* form, although slowly-biodegradable soluble organic matter may be present. The slowly-biodegradable matter consists of relatively complex molecules that are not directly used by the bacteria. For this to occur, the conversion into soluble matter is necessary, through the action of extracellular enzymes. This conversion mechanism, called *hydrolysis*, does not involve the use of energy, but results in the delay in the consumption of the organic matter (see Chapter 1).

3.1.3 Characterisation of the nitrogenous matter

The first major division in the nitrogenous matter that enters a sewage treatment works is by its organic state: the nitrogenous matter may be (a) *inorganic* or (b) *organic*.

- The **inorganic nitrogen** is represented by *ammonia*, either in its free form (NH_3) or in its ionised form (NH_4^+). Ammonia is present in the influent sewage because the hydrolysis and ammonification reactions, described below, have already started in the sewerage system.
- The **organic nitrogen** is divided in a similar form to the carbonaceous matter, as a function of the biodegradability: (a) *inert* and (b) *biodegradable*.
  - *Inert*. The inert fraction is divided into two fractions, according to the physical state:
    - *Soluble*. This part is usually negligible and does not need to be considered.
    - *Particulate*. This part is associated with the non-biodegradable carbonaceous organic matter, being involved by the biomass and removed with the excess sludge.
  - *Biodegradable*. The biodegradable fraction can be subdivided into the following three components:
    - *Rapidly biodegradable*. The rapidly-biodegradable organic nitrogenous matter is in a *soluble* form and is converted by heterotrophic bacteria into ammonia, through the process of *ammonification*.
    - *Slowly biodegradable*. The slowly-biodegradable organic nitrogenous matter is in a *particulate* form, being converted into a soluble

form (rapidly biodegradable) through *hydrolysis*. This hydrolysis occurs in parallel with the hydrolysis of the carbonaceous matter.

– *Ammonia*. Ammonia (inorganic nitrogen) results from the hydrolysis and ammonification processes described above. Ammonia is used by heterotrophic and autotrophic bacteria.

3.1.4 Participation of the biomass

From the above, it is assumed that (a) the *carbonaceous matter* used directly by the bacteria is present in a *soluble* form (rapidly biodegradable), and that (b) the *nitrogenous matter* used directly by the bacteria will be present in the form of *ammonia*.

The biomass present in biological treatment systems can be divided into the following groups, as a function of its *viability*: (a) *active biomass* and (b) *inert residue*.

- The **inert residue** is formed through the decay of the biomass involved in the sewage treatment. The biomass decay can take place according to various mechanisms, which include endogenous metabolism, death, predation and others. As a result, slowly-degradable products are generated, as well as particulated products, inert to biological attack.
- The **active biomass** is that responsible for the biological degradation of the compounds. Depending on the carbon source, the biomass can be divided into (a) *heterotrophic* and (b) *autotrophic* (see Chapter 1):
  - *Active heterotrophic biomass*. The carbon source for the heterotrophic organisms is the carbonaceous organic matter. The heterotrophic biomass uses the rapidly biodegradable carbonaceous matter (soluble). Part of the energy associated with these molecules is incorporated into the biomass, while the remainder is used to supply energy for synthesis. In aerobic treatment, the growth of the heterotrophic biomass is possible in aerobic (using oxygen as an electron acceptor – see Chapter 1) or anoxic (absence of oxygen, by using the nitrate as an electron acceptor) conditions, but is very low in anaerobic conditions (absence of oxygen or nitrate). The heterotrophic bacteria use nitrogen in the form of ammonia for synthesis (in aerobic and anoxic conditions) and nitrogen in the form of nitrate as an electron acceptor (in anoxic conditions). The decay of the heterotrophic biomass generates, besides the inert residue, slowly degradable carbonaceous and nitrogenous matter. These subsequently need to undergo the hydrolysis process, to be converted into rapidly-biodegradable matter, which can be used again by the heterotrophic and autotrophic biomass.

- *Active autotrophic biomass*. The carbon source for the autotrophic organisms is carbon dioxide. The autotrophic biomass uses ammonia as the energy source (they are chemoautotrophic organisms, that is, they use inorganic matter as energy source). In aerobic conditions, these bacteria use ammonia in the nitrification process, in which ammonia is converted into nitrite and then nitrate. Similarly to the heterotrophic organisms, the decay of the autotrophic biomass also generates, besides an inert residue, slowly-degradable carbonaceous and nitrogenous matter. These subsequently need to undergo hydrolysis, to be converted to rapidly-biodegradable matter, which can be used again by the heterotrophic and autotrophic biomass.

3.1.5 Representation of the biomass and the substrate

3.1.5.1 Representation of the biomass

Due to the difficulty in characterising the biological solids and the substrate according to the above concepts, most of the mathematical models introduce simplifications in their representation. Such simplified representations are described in the present section.

The unit of mass of the microbial cells is normally expressed in terms of *suspended solids* (SS), since the biomass consists of solids that are suspended in the reactor (in the case of dispersed growth). However, not all the solids mass participates in the conversion of the organic substrate, as there is an inorganic fraction that does not play an active role in biological treatment. Therefore, the biomass is frequently expressed in terms of *volatile suspended solids* (VSS). These represent the organic fraction of the biomass – the organic matter can be volatised, that is, converted into gas by combustion (oxidation).

However, as mentioned, not all the organic fraction of the biomass is really active (Eckenfelder, 1980; Marais and Ekama, 1976; Grady and Lim, 1980; IAWPRC, 1987). Thus, the volatile suspended solids can be divided into an *active* and an *inactive* fraction. The active fraction is that which has the real participation in the conversion of the substrate. The main limitation of the use of the active solids in the design and operational control of a treatment plant relates to the difficulty in their determination. There are indirect processes, based on DNA, ATP, proteins, and others, but none compares to the simplicity of the direct determination of volatile suspended solids.

Besides considering the biomass activity, the solids can also be interpreted with relation to their biodegradability. Not all the volatile suspended solids are biodegradable, and there is a *biodegradable* and a *non-biodegradable* fraction.

In summary, the following distribution is frequently adopted for the suspended solids in a reactor:

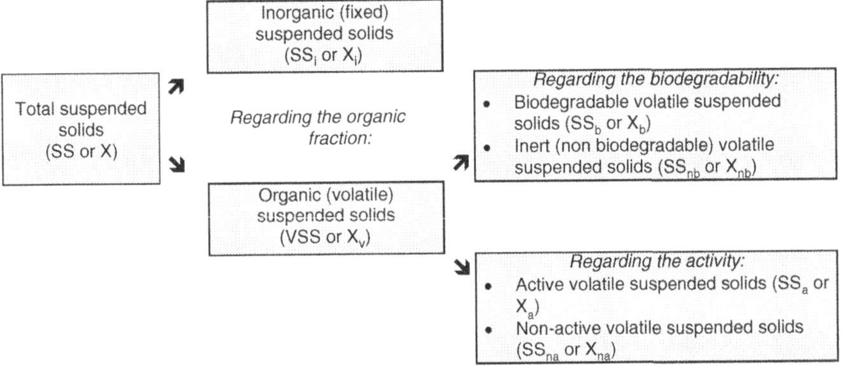

3.1.5.2 Representation of the organic matter

As mentioned, the organic matter can be considered as having a *soluble* fraction, corresponding to the dissolved organic solids (most being rapidly biodegradable), and a *suspended* or *particulate* fraction, relative to the suspended solids (slowly biodegradable). In terms of carbonaceous matter, the present text adopts BOD_5 or COD as variables representing the substrate. In order to make the text more applicable, the treatment processes that have been traditionally designed using BOD_5 maintain it as the basic variable. On the other hand, the more recent processes that have been using COD for design are also described in terms of COD.

As a result, the following variables are adopted in terms of the representation of the substrate (carbonaceous matter):

- *Influent substrate S_o (influent BOD_5 or COD)*. Represents the *total* **BOD_5** (soluble BOD + particulate BOD) or *total* **COD** (soluble COD + particulate COD) influent to the biological reactor.

  Even in systems with primary sedimentation, around 1/3 of the suspended solids are not removed in this stage and enter the biological reactor. In the reactor, suspended solids are adsorbed by the biomass and are converted into soluble solids by hydrolysis mechanisms, after which they undergo the conversion reactions. Therefore, in the influent to the reactor, the soluble substrate as well as the particulate substrate must be computed as the influent substrate to be removed.

- Effluent substrate *S (effluent BOD_5 or COD)*. Represents the effluent *soluble* **BOD_5** or *soluble* **COD** from the reactor.

  Even though the effluent from the reactor could contain a high concentration of suspended solids (biological solids that compose the biomass), these solids are largely removed in the subsequent settling stage, when existent (e.g. secondary sedimentation tank or sedimentation lagoons). In the design of a reactor that receives recycled solids, there is no point in computing the effluent *total BOD* or *COD* from it, because it can be occasionally larger than the influent BOD or COD, due to the high concentration of particulate organic matter represented by the microbial population. The quality of the final effluent from the treatment plant depends on the (a) *soluble BOD or COD:* reactor performance; (b) *particulate BOD or COD:* performance of the final settling unit (when existent) or the concentration of the effluent solids from the reactor (when there is no final settling unit).

3.2 CONVERSION PROCESSES OF THE CARBONACEOUS AND NITROGENOUS MATTERS

3.2.1 Conversion of the carbonaceous matter

3.2.1.1 Aerobic conversion

The general equation of aerobic respiration can be expressed as:

$$\underset{\text{organic matter}}{C_6H_{12}O_6} + 6\,O_2 \longrightarrow 6\,CO_2 + 6\,H_2O + \text{Energy} \tag{3.1}$$

This equation is general and simplified, since, in reality, there are various intermediate steps. The composition of the organic matter is simplified and, in this case, the molecular formula of the glucose is assumed as a representation of the carbonaceous organic matter. By analysing the reaction, the following aspects can be highlighted, all important in sewage treatment (Branco, 1976):

* Stabilisation of the organic matter (conversion to inert products, such as carbon dioxide and water).
* Utilisation of oxygen.
* Production of carbon dioxide.
* Release of energy.

Equation 3.1 can be expressed in a generic way by an organic compound with molecular formula $C_xH_yO_z$, which permits the calculation of the oxygen consumption and the production of carbon dioxide (van Haandel and Lettinga, 1994):

$$C_xH_yO_z + \frac{1}{4}(4x + y - 2z)\,O_2 \rightarrow x\,CO_2 + \frac{y}{2}H_2O \tag{3.2}$$

As mentioned, Equations 3.1 and 3.2 are generic, covering only the oxidation of the *carbonaceous* organic matter. Other elements (such as nitrogen, phosphorus, potassium, etc.) are frequently part of the composition of the organic matter, which is still able to undergo biochemical oxidation.

The main agents responsible for the aerobic stabilisation of the carbonaceous matter contained in the sewage are the decomposing organisms, which are mainly represented by aerobic and facultative heterotrophic bacteria.

3.2.1.2 Anaerobic conversion

The conversion of the carbonaceous matter under anaerobic conditions follows the equation below:

$$\underset{\text{organic matter}}{C_6H_{12}O_6} \longrightarrow 3\,CH_4 + 3\,CO_2 + \text{Energy} \qquad (3.3)$$

This equation is also general and simplified, and represents only the final product from the intermediate stages. The following aspects can be highlighted in the equation:

- Non-exclusivity of the oxidation. The carbon of CO_2 is present in its highest state of oxidation (+4). However, the opposite occurs with CH_4, in which the carbon is in its most reduced state (−4), subsequently being able to be oxidised (for example, by combustion – methane is inflammable).
- No utilisation of oxygen.
- Production of methane and carbon dioxide.
- Release of energy (less than in aerobic respiration).

The organic matter was only converted to a more oxidised form (CO_2) and another more reduced form (CH_4). However, most of the CH_4 is released to the gaseous phase, which then leads to an effective removal of the organic matter.

Equation 3.3 can be expressed in a generic way for an organic compound $C_xH_yO_z$ as (van Haandel and Lettinga, 1994):

$$C_xH_yO_z + \frac{4x - y - 2z}{4}H_2O \rightarrow \frac{4x - y + 2z}{8}CO_2 + \frac{4x + y - 2z}{8}CH_4 \qquad (3.4)$$

The anaerobic conversion occurs in two stages:

- **Acidogenic phase**: conversion of the organic matter into organic acids by *acidogenic* organisms (acid-forming organisms). In this stage, there is only the conversion of organic matter, but no removal.
- **Methanogenic phase**: conversion of the organic acids into methane, carbon dioxide and water by methanogenic organisms (methane-forming organisms). The organic matter is converted again, but because CH_4 is transferred to the atmosphere, there is the removal of the organic matter.

Before the acidogenesis stage, the complex organic compounds (carbohydrates, proteins, and lipids) need to be converted into simple organic compounds, through the mechanism of **hydrolysis**.

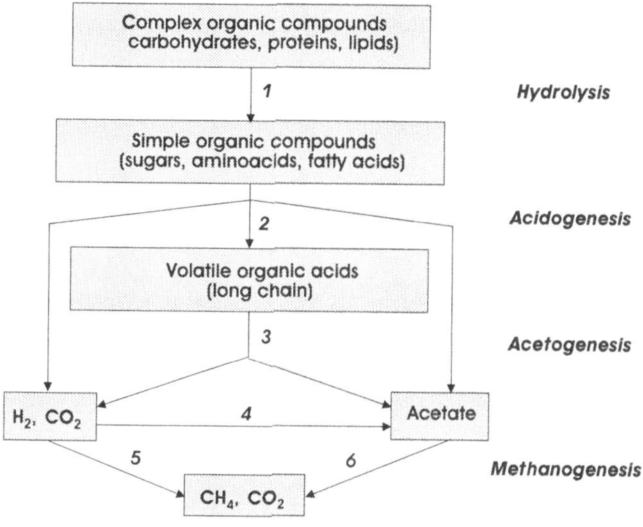

Figure 3.2.   Metabolic sequences and microbial groups involved in anaerobic digestion (Chernicharo, 1995; Lubberding, 1995)

Figure 3.2 illustrates the sequence of stages involved in the anaerobic digestion of the organic matter.

3.2.2 Conversion of nitrogenous matter

3.2.2.1 Oxidation of ammonia (nitrification)

In domestic sewage, the organic nitrogen is converted into ammonia, through the process of **ammonification**. This process does not change the quantity of nitrogen (TKN) in the wastewater, has no consumption of oxygen, and starts in the sewerage system itself, continuing in the primary and biological treatment units. In the end of the treatment, the quantity of organic nitrogen is small.

An important oxidation reaction that occurs in some wastewater treatment processes is the **nitrification**, in which the ammonia is transformed into nitrites and these nitrites into nitrates. Only some treatment processes are able to support a significant nitrification, because of their capacity of maintaining sufficient concentrations of the nitrifying bacteria.

The microorganisms involved in these processes are chemoautotrophs, for which carbon dioxide is the main source of carbon, and the energy is obtained through the oxidation of an inorganic substrate, such as ammonia, to mineralised forms.

The transformation of ammonia into nitrites is done by bacteria, such as those from the genus *Nitrosomonas*, according to the following reaction:

$$2NH_4^+ + 3O_2 \xrightarrow{\textit{Nitrosomonas}} 2NO_2^- + 4H^+ + 2H_2O \qquad (3.5)$$

The oxidation of the nitrites to nitrates occurs mainly by the action of bacteria, such as those from the genus *Nitrobacter*, according to:

$$2\,NO_2^- + O_2 \xrightarrow{\textit{Nitrobacter}} 2\,NO_3^- \qquad (3.6)$$

The global reaction of nitrification is the sum of Equations 3.5 and 3.6:

$$NH_4^+ + 2\,O_2 \longrightarrow NO_3^- + 2\,H^+ + H_2O \qquad (3.7)$$

In the reactions 3.5 and 3.6 (as well as in the global reaction 3.7), the following points should be noted:

- Consumption of free oxygen. This consumption is called *nitrogenous demand*.
- Release of H^+, consuming the alkalinity of the medium and possibly reducing the pH.

Figure 3.3 shows a typical distribution of the nitrogen compounds in a treatment system after ammonification and the subsequent nitrification. The oxidised forms of nitrogen (nitrites and nitrates) are collectively called NO_x. It is seen that with nitrification there is *no removal of nitrogen* (total nitrogen remains the same), but only conversion of the nitrogen forms.

3.2.2.2 Reduction of nitrate (denitrification)

In anoxic conditions (absence of oxygen, but in the presence of nitrates), the nitrates are used by heterotrophic organisms as an electron acceptor instead of oxygen. In this process, called **denitrification**, *nitrate* is reduced to *nitrogen gas*, according to the following reaction:

$$2\,NO_3^- + 2\,H^+ \longrightarrow N_2 + 2.5\,O_2 + H_2O \qquad (3.8)$$

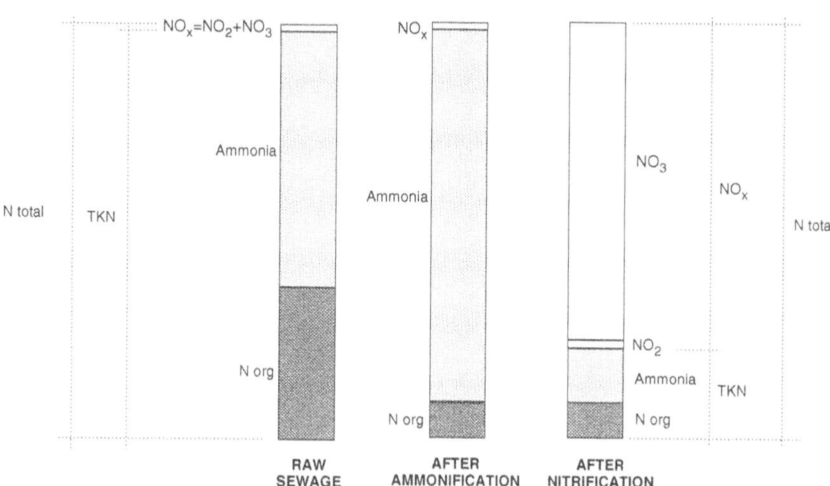

Figure 3.3.   Distribution of nitrogen in a treatment system with nitrification

With the denitrification reaction, the following points should be noted:

- Economy of oxygen (the organic matter can be stabilised in the absence of oxygen)
- Consumption of H^+, implying an economy of the alkalinity and an increase in the buffer capacity of the medium

Figure 3.4 shows a typical distribution of the nitrogen forms in a treatment system with nitrification and denitrification. It is seen that, besides the conversion in the forms of nitrogen, there is also the *removal of nitrogen* (total nitrogen is decreased). In other words, denitrification leads to an effective removal of nitrogen from the liquid, corresponding to the nitrate that is converted to nitrogen gas, which escapes to the atmosphere.

3.3 TIME PROGRESS OF THE BIOCHEMICAL OXIDATION OF THE CARBONACEOUS MATTER

In a simplified manner and neglecting intermediate mechanisms, it can be said that the *aerobic* reactions for the stabilisation of the organic matter occur, in a closed system (such as the bottle used in the BOD test), in a sequence in which the two following main mechanisms are predominant (Eckenfelder, 1980):

- *Initial stage: predominance of synthesis (anabolism)*

   At the beginning, the organic matter present in the wastewater is used by the microorganisms for their metabolic activities of growth and energy

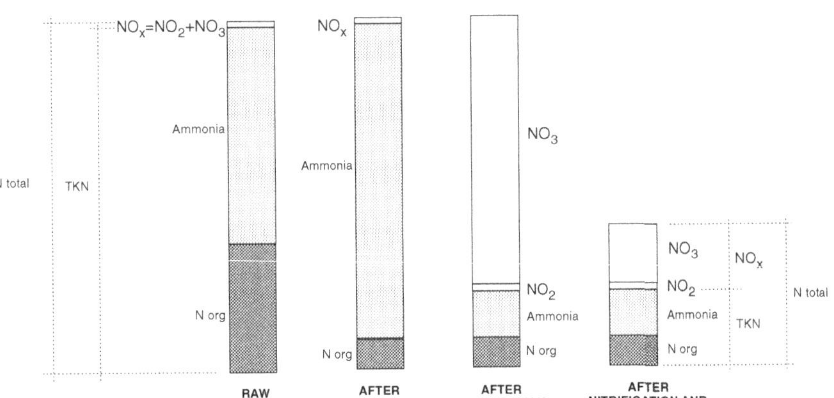

Figure 3.4.   Distribution of nitrogen in a treatment system with nitrification and denitrification

conversion, therefore prevailing the activities related to *synthesis*. This phase results in oxygen consumption and in the increase in the microorganisms population, and can be represented by the generic equation (Hanisch, 1980):

$$8\,CH_2O + NH_3 + 3\,O_2 \longrightarrow C_5H_7NO_2 + 3\,CO_2 + 6\,H_2O + Energy \qquad (3.9)$$

organic matter   cellular matter

In this equation, CH_2O represents the organic matter, in the same way that the equivalent formulation $C_6H_{12}O_6$ was also used to represent organic matter in Item 3.2.1. The cellular matter is expressed by the following empirical formula $C_5H_7NO_2$ (see Section 1.7.2).

• *Subsequent stage: predominance of endogenous respiration (catabolism)*

When the organic matter originally present in the wastewater is mostly removed, there is the predominance of the mechanisms of the second phase of oxidation. At the start of this phase, the microorganisms' population is at their maximum and, due to the low availability of substrate in the medium, the main food source becomes their own cellular protoplasm. Therefore, in this phase there is the predominance of auto-oxidation mechanisms or *endogenous respiration* (see Section 3.4, about bacterial growth). The simplified representative equation of this phase is:

$$C_5H_7NO_2 + 5\,O_2 \longrightarrow 5\,CO_2 + NH_3 + 2\,H_2O + Energy \qquad (3.10)$$

cellular matter

Naturally in a system with mixed cultures, such as the reactors used in biological wastewater treatment, there are microorganisms with different growth and decay rates. Consequently, some microorganism species can be in one stage of synthesis or endogenous respiration, while other species are in earlier or later phases. The representation above regards only average conditions for heterotrophic microorganisms in the reactor.

The total oxygen consumed in both phases is defined as the *ultimate oxygen demand* (BOD_u). The addition of equations 3.9 and 3.10 leads to the simplified equation for the oxidation of the organic matter (identical to Equation 3.1):

$$\boxed{CH_2O + O_2 \longrightarrow CO_2 + H_2O + \text{Energy}} \qquad (3.11)$$

The value of the theoretical oxygen demand (ThOD) based on the stoichiometric relations of Equation 3.11 differs, in a certain way, from what is found for the ultimate demand, being, in reality, a little higher. This is because in the endogenous phase, when a bacteria dies, it becomes food for other bacteria, and thus a subsequent transformation to carbon dioxide, water and cellular material occurs. The living as well as the dead bacteria serve as food for higher organisms, such as protozoa. In each transformation, a new oxidation occurs, but in the general balance, a certain fraction of the organic matter, resistant to biological attack, remains. This fraction is the one responsible for the deviation between the values of the theoretical and the ultimate oxygen demand (Sawyer and Mc Carty, 1978).

The removal and oxidation of the organic matter present in the wastewater (first phase) normally has a duration of one to two days. The total oxidation of the cellular mass will take a very long time, but, in practical terms for domestic sewage, it can be considered complete around 20 days. The reaction rate in the assimilation phase is several times higher than in the second phase, of endogenous respiration (Eckenfelder, 1980).

Figure 3.5 presents the curves of the accumulated oxygen consumption, substrate concentration and bacteria concentration as a function of time

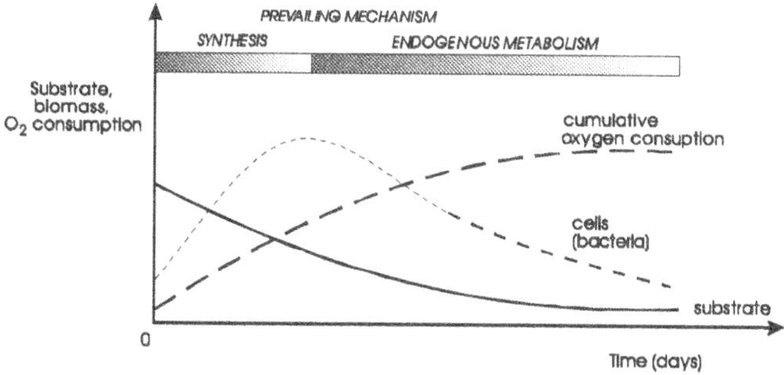

Figure 3.5.   Oxidation of the carbonaceous matter along time

3.4 PRINCIPLES OF BACTERIAL GROWTH

3.4.1 Synthesis and endogenous respiration

As already mentioned, heterotrophic organisms use organic matter as a form of condensed energy that is necessary for their various metabolic processes, including growth and reproduction. With the use of oxygen (aerobic conditions – Equation 3.1) or another electron acceptor (e.g. nitrate, in anoxic conditions – Equation 3.8), these organisms oxidise the organic matter, with the production of more cellular matter (growth and reproduction) and energy release. This is the **synthesis** phase.

If the substrate available in the medium starts to become scarce, such as in sewage treatment, in which organic matter is progressively removed, the microorganisms need to find other organic matter or condensed energy sources. The main substrate directly available is their own cellular protoplasm, which the cells start to use according to Equation 3.10. In this stage, the balance is negative, that is, there is a reduction in the cellular matter or the bacterial concentration in the medium, characterising the **endogenous respiration** stage.

It is thus seen that there is a close relation between the substrate concentration in the medium, or the available food, and the microorganisms' population. When the availability of organic matter is sufficient, the bacteria are in a growing phase, and when it becomes insufficient, the bacteria enter a decreasing stage. This consideration is of large importance in sewage treatment, in which systems can be designed to operate with a high or low organic matter supply for the bacteria.

Besides, the form in which the two phases are located in the biological reactor depends on its hydraulic configuration (see Chapter 2). In a *plug-flow* reactor, the reaction time is associated with the physical location in the reactor. Hence, the sequencing between the two phases can take place along the inlet and outlet of the reactor. In a *complete-mix* reactor, the concentration of the substrate and the bacteria are the same at any point in the reactor. Thus, the relative predominance of one phase or another will depend on the prevalent concentration of the substrate in the reactor. If it is high, the synthesis phase prevails in the reactor as a whole. However, if the substrate concentration is low, the balance favours the mechanisms of endogenous respiration.

A simplified scheme of the heterotrophic bacterial metabolism is presented in Figure 3.6.

3.4.2 Bacterial-growth curve

The main reproduction mode for bacteria is by binary fission, in which, when the cell reaches a certain size, it splits into two cells, which will subsequently generate four new cells and so on. Thus, after n divisions the number of cells formed is 2^n. Assuming a typical generation time of 20 minutes, a population growth without limiting factors could lead to 2^{144} bacteria after 48 hours. This would correspond to a weight approximately 4000 times greater than the weight of the earth (La Riviére,

HETEROTROPHIC BACTERIAL METABOLISM

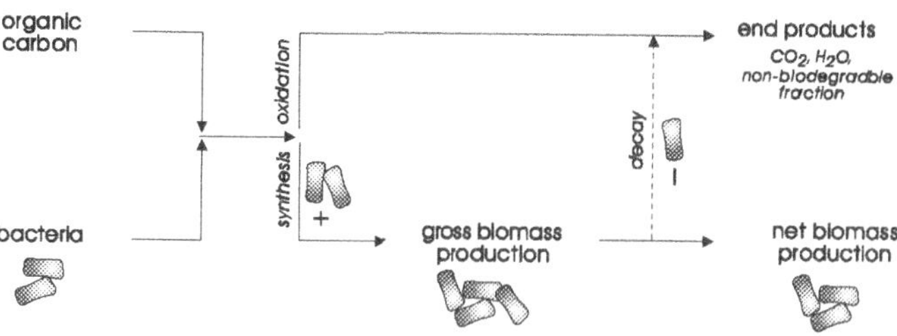

Figure 3.6.　A simplified scheme of the heterotrophic bacterial metabolism

TIME PROGRESS OF BACTERIAL GROWTH

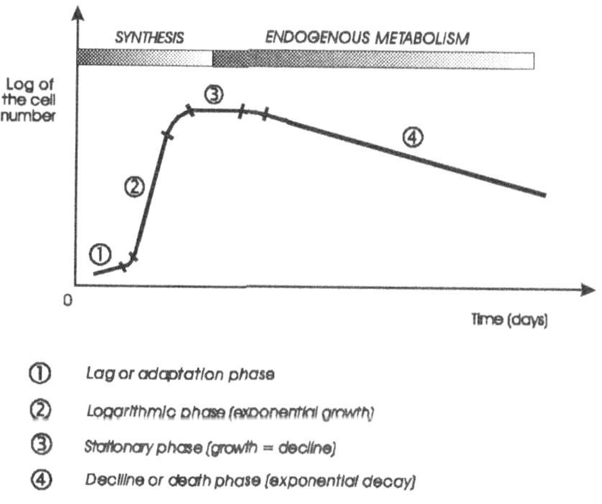

① 　Lag or adaptation phase
② 　Logarithmic phase (exponential growth)
③ 　Stationary phase (growth = decline)
④ 　Decline or death phase (exponential decay)

Figure 3.7.　Typical bacterial-growth curve

1980). Naturally, in practice the growth is soon restricted due to the exhaustion of the nutrients in the medium.

When inoculating a liquid volume with a certain initial quantity of bacterial cells and a limited quantity of substrate, the number of bacteria will progress with time according to the typical bacterial growth-curve, expressed in Figure 3.7 (vertical logarithmic scale).

- **Lag phase**. The lag phase is a period for enzymatic adaptation of the bacteria to the new substrate supplied (von Sperling, 1983a). This phase can be reduced in the case of typical domestic sewage, in which the bacteria have already acquired the necessary enzymatic equipment.

- **Exponential-growth phase**. In the exponential growth phase the cells divide themselves at a constant rate. Plotted on a logarithmic scale, the number of cells grows linearly, justifying the alternative designation of *logarithmic phase*. There is an excess of the substrate in the medium, allowing the growth rate to reach its maximum, with the only limitation by the microorganisms' capacity to process the substrate. In parallel with the maximum growth rate, there is also the maximum substrate removal rate.
- **Stationary phase**. The stationary phase is when the food starts to be scarce in the medium, and the bacterial growth rate is equal to the death rate. Therefore, the number of cells is maintained temporally constant.
- **Decline or decay phase.** In the decline or decay phase, the availability of the substrate in the medium is reduced. In these conditions, *endogenous respiration* prevails, and the bacteria are forced to use their own cellular protoplasm as a substrate source. The dying cells allow their nutrients to diffuse into the medium, serving as food to other cells. The death rate is exponential and constant, leading to a straight line on the logarithmic scale.

As already mentioned, it is important to emphasise that this representation of the growth regards a *single* population of microorganisms growing at the expense of a *single* type of substrate. In reality, in the biological reactor of a sewage treatment works, there is a variety of microorganisms metabolising a variety of compounds. Hence, there will be an overlapping of various curves of different forms and types, developing at different times. This interaction characterises the ecology of wastewater treatment, covered in Chapter 1.

The design and operation of a sewage treatment plant uses these concepts of bacterial growth to place the operation inside a desired range. A generalisation is difficult due to the large variety of microorganisms and substrates that occur in practice, but the following tendencies can be observed:

- **Very high loading systems**. In the exponential growth phase, the substrate availability is high. This indicates that the concentration of the substrate (e.g. BOD) in the effluent will also be somewhat high. Thus, the majority of sewage treatment systems do not operate in this phase.
- **High loading systems** (e.g. conventional activated sludge, high rate trickling filters). The concentration of substrate in the effluent is lower, but the cellular mass has a high organic fraction, requiring the separate stabilisation of the excess sludge. Due to the high load, the volume required for the reactor is smaller than in the low loading systems.
- **Low loading systems** (e.g. activated sludge systems of the extended aeration variant, low rate trickling filters). The reasoning of these systems is to supply a minimum quantity of substrate to the organisms, in order to stimulate endogenous respiration. This leads to a self metabolism of the microorganisms, that is, they undergo a digestion of the cellular mass in the reactor itself. Besides the partial stabilisation of the cellular mass, the concentration of the substrate in the effluent is very low. The volume required for the reactor is larger than for the high loading systems.

3.4.3 Kinetics of bacterial growth

3.4.3.1 Specific gross bacterial growth

The bacterial growth can be expressed as a function of the bacteria concentration at a given time in the reactor. The *net* growth rate is equal to the gross growth rate minus the bacterial decay rate.

The growth rate of a bacterial population is a function of its number, mass or concentration at a given time. Mathematically, this relation can be expressed as:

$$\boxed{\frac{dX}{dt} = \mu.X} \tag{3.12}$$

where:

X = concentration of the microorganisms in the reactor, SS or VSS (g/m^3)
μ = specific growth rate (d^{-1})
t = time (d)

This formula, if integrated, assumes an exponential form, which, when plotted on a logarithmic scale, results in a straight line. This is the logarithmic phase shown in Figure 3.7.

The growth rate, such as expressed by Equation 3.12, is for a growth without limitation of substrate. However, it was seen in the previous sections that bacterial growth is a function of the availability of the substrate in the medium. When the substrate is present at a low concentration, the growth rate is proportionally low. In sewage treatment, the carbonaceous matter is usually the limiting growth factor.

The specific growth rate μ must be therefore expressed as a function of the substrate concentration. Monod, in his classic studies with bacterial cultures, presented this relation according to the following empirical formula:

$$\boxed{\mu = \mu_{max} \cdot \frac{S}{K_s + S}} \tag{3.13}$$

where:

μ_{max} = maximum specific growth rate (d^{-1})
S = concentration of the limiting substrate or nutrient (g/m^3)
K_s = half-saturation coefficient, which is defined as the substrate concentration for which $\mu = \mu_{max}/2$ (g/m^3)

Figure 3.8 shows a graphic representation of this equation. The *limiting nutrient* (S) is the one that, in case its concentration is reduced, will lead to a decrease in the population growth rate (as indicated in the figure through the reduction of μ). Conversely, if the concentration of S starts to increase, the population will consequently increase. However, if S continues to increase, there will be a point in

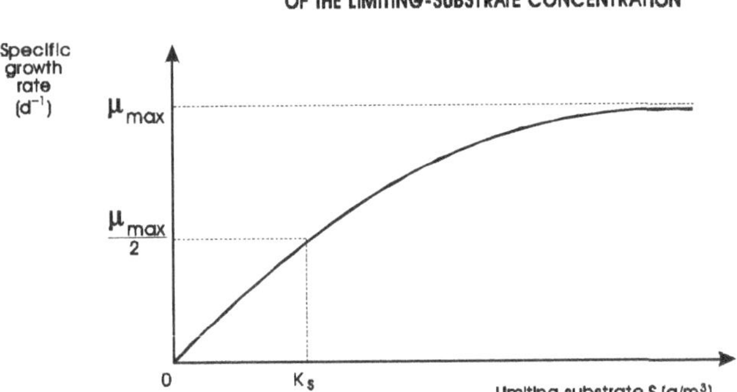

Figure 3.8.   Specific growth rate as a function of the concentration of the limiting substrate

which it will be in excess in the medium, not being anymore the limiting factor for the population growth. In these conditions, another nutrient will probably start to control the growth, becoming the new limiting factor. This explains why μ tends to a maximum value, expressed by μ_{max}. At this point, even if the concentration S increases, μ will not increase further, since it is no longer limited by S. The basis of the Monod formulation are analysed in Chapter 1.

The interpretation of the half-saturation coefficient K_s is that, when the substrate concentration in the medium is equal to K_s ($K_s = S$), the term $S/(K_s + S)$ of Equation 3.13 becomes equal to $^1/_2$. Thus, the growth rate μ becomes equal to half the maximum rate ($\mu_{max}/2$). To compare different substrates, the value of K_s gives an indication of the *non-affinity* of the microorganisms for each substrate: the greater the value of K_s, the lower the growth rate μ or, else, the lower the affinity of the biomass to the substrate. To obtain high substrate removals in sewage treatment, it is desirable that the substrate has low values of K_s.

In the case of the heterotrophic bacteria involved in sewage treatment, the limiting substrate is usually organic carbon or, in other words, BOD or COD. This is because the reactors work with low organic carbon concentrations to produce an effluent with a low BOD concentration.

Under other conditions, there can be other limiting nutrients, or even a composition of two or more of them. This is the case of the growth rate of the nitrifying organisms. Due to the fact that their growth rate is small, either with low ammonia values as well as with low dissolved oxygen values, the Monod relation can be expressed as a double-inhibition function. Hence, instead of only one term $S/(K_s + S)$, there is the product of two terms $[S_1/(K_{s1} + S_1)] \cdot [S_2/(K_{s2} + S_2)]$, in which S_1 and S_2 are the concentrations of the two limiting factors (in this case, ammonia and oxygen).

In the laboratory, the curve in Figure 3.8 (or a transformation of it) can be constructed, and the values of K_s and μ_{max} can be extracted. In domestic sewage treatment, values of K_s and μ_{max} in the following ranges have been reported:

- *Aerobic treatment (Metcalf & Eddy, 1991):*

$$\mu_{max} = 1.2 \text{ to } 6 \text{ d}^{-1}$$

$$K_s = 25 \text{ to } 100 \text{ mg BOD}_5/\text{l}$$
$$\text{or}$$
$$K_s = 15 \text{ to } 70 \text{ mgCOD}/\text{l}$$

- *Anaerobic treatment (van Haandel and Lettinga, 1994; Chernicharo, 1997):*

$$\mu_{max} = 2.0 \text{ d}^{-1} \text{ (acidogenic organisms)}$$
$$\mu_{max} = 0.4 \text{ d}^{-1} \text{ (methanogenic organisms)}$$
$$\mu_{max} = 0.4 \text{ d}^{-1} \text{ (combined biomass)}$$

$$K_s \approx 200 \text{ mgCOD}/\text{l (acidogenic organisms)}$$
$$K_s \approx 50 \text{ mgCOD}/\text{l (methanogenic organisms)}$$

Certain types of organisms could of course have different coefficients from these global values.

The Monod equation has the same form as the Michaelis–Menten equation for enzymatic relations (see Chapter 1). However, while the latter is based on theoretical principles, the Monod relation is essentially empirical. Another aspect is that the Monod equation was derived for a single organism metabolising a single substrate. However, in wastewater treatment this assumption is not valid, since there is a *multiple population* assimilating a *multiple substrate*. Due to these aspects, the Monod relation has been the target of criticism from the specialised literature. However, a more satisfactory relation has not yet been developed, and the Monod equation maintains its importance, being adopted in practically all mathematical models of biological wastewater treatment.

A great advantage of the Monod equation resides in its structure that permits the representation in a continuous form of the range of variation between the extremes of lack and abundance of nutrients in the medium. Therefore, depending on the value of S, the Monod equation can represent approximately the kinetics of zero and first orders, as well as the transition between them. In the case of a substrate removal reaction, when its concentration is still high and not limiting, the global removal rate approaches the *zero-order* kinetics. With the consumption of the substrate, the reaction starts to decrease, characterising a transition or mixed-order region. When the substrate concentration is very low, the reaction rate starts to be limited by its low availability in the medium. In these conditions, the reaction kinetics approach *first order*. These two situations occur depending on the relative values of S and K_s, as described below:

- Relative substrate concentration: **high**

> $S \gg K_s$ •   *reaction approximately zero order*
>   •   *growth rate μ independent of S*

When the substrate concentration is much higher than the value of K_s, K_s can be neglected in the denominator of Equation 3.13, which is reduced to:

$$\mu = \mu_{max} \tag{3.14}$$

In these conditions, the growth rate μ is constant and equal to the maximum rate μ_{max}. The reaction follows a *zero-order kinetics*, in which the reaction rate is independent from the substrate concentration. In domestic sewage treatment, this situation tends to occur at the head of a plug-flow reactor, where the substrate concentration is still high.

- Relative substrate concentration: **low**

> $S \ll K_s$ •   *reaction approximately first order*
>   •   *growth rate μ is dependent on S (directly proportional)*

When the substrate concentration is much lower than the value of K_s, S can be neglected in the denominator of Equation 3.13, which is reduced to:

$$\mu = \mu_{max} \cdot \frac{S}{K_s} \tag{3.15}$$

As μ_{max} and K_s are constants, the term (μ_{max}/K_s) is also a constant, and can be substituted by a new constant K. Consequently, Equation 3.15 is reduced to:

$$\mu = K.S \tag{3.16}$$

In this situation, the growth rate is proportional to the substrate concentration. The reaction follows *first-order kinetics*. This situation is typical in the treatment of domestic sewage in a complete-mix reactor in which the substrate concentration in the medium is low due to the requirements of low values in the effluent.

Figure 3.9 presents the two extreme situations that represent the zero and first order kinetics.

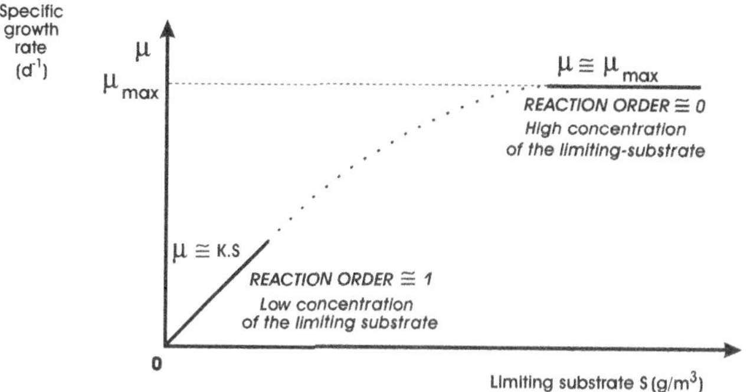

Figure 3.9.   Extreme conditions in the saturation reaction (Monod kinetics)

Example 3.1

Express μ as a function of μ_{max} for the following conditions:

- domestic sewage; S = 300 mg/L (adopt K_s = 40 mg/L)
- domestic sewage; S = 10 mg/L (adopt K_s = 40 mg/L)
- glucose; S = 10 mg/L (adopt K_s = 0.2 mg/L)

Solution:

a)   Domestic sewage (S = 300 mg/L)
     From Equation 3.13:

$$\mu = \mu_{max} \cdot \frac{S}{K_s + S} = \mu_{max} \cdot \frac{300}{40 + 300} = 0.88 \mu_{max}$$

Hence, $\mu = 0.88\ \mu_{max}$

In these conditions, in which S is large in comparison with K_s, the growth rate μ is close to μ_{max}. There is a great availability of the limiting nutrient and the population presents a high growth rate. The reaction is approximately zero order. This situation is not very frequent in the treatment of domestic sewage and occurs at the head of a plug-flow reactor, where the substrate concentration is still high.

b)   Domestic sewage (S = 10 mg/L)

$$\mu = \mu_{max} \cdot \frac{S}{K_s + S} = \mu_{max} \cdot \frac{10}{40 + 10} = 0.20\ \mu_{max}$$

Thus, $\mu = 0.20\ \mu_{max}$

Example 3.1 (Continued)

As S is small in comparison with K_s, the growth rate is much lower than μ_{max}. This indicates that there is not much availability of the limiting nutrient in the medium. This situation is typical in a complete-mix reactor treating domestic sewage, in which the substrate is completely homogenised and is present at a concentration equal to the effluent one.

c)   Glucose (S = 10 mg/L)

$$\mu = \mu_{max} \cdot \frac{S}{K_s + S} = \mu_{max} \cdot \frac{10}{0.2 + 10} = 0.98\,\mu_{max}$$

Hence, $\mu = 0.98\,\mu_{max}$

The concentration of S is the same as in item b, but since in the case of glucose K_s is very low (high affinity), the denominator of the expression is practically equal to S and, as a result, μ is almost equal to μ_{max}. Consequently, there is a high availability of the substrate and the growth rate is very close to the maximum.

3.4.3.2 Bacterial decay

The relations presented in the previous section correspond to the gross biomass growth. However, since the bacteria stay in the treatment systems for more than one or two days, there is also the endogenous metabolism stage. This implies that part of the cellular matter is destroyed by means of some of the mechanisms active in the endogenous respiration stage. To obtain the *net* growth rate, this loss should be discounted, which is also a function of the concentration or bacterial mass. For accuracy, only the *biodegradable fraction* of the biomass should be considered, since there is also an inert, non-biodegradable organic fraction, not subject to bacterial decay. For simplicity, the total VSS are considered in most of this chapter, and not the biodegradable VSS. In the chapters related to the activated sludge system, this concept is deepened, and the concept of the biodegradable fraction is used.

The decay rate can be expressed as a first-order reaction:

$$\frac{dX}{dt} = -K_d \cdot X \tag{3.17}$$

where:

K_d = endogenous respiration coefficient, or bacterial decay coefficient (d^{-1})

For typical domestic sewage, K_d varies in the following ranges:

- *Aerobic treatment:*

$K_d = 0.04$ to 0.10 mgVSS/mgVSS.d (base: BOD_5) (Metcalf & Eddy, 1991; von Sperling, 1997)

or

$K_d = 0.05$ to 0.12 mgVSS/mgVSS.d (base: COD) (EPA, 1993; Orhon and Artan, 1994)

- *Anaerobic treatment:*
  The values available in the literature appear to be not very reliable (Lettinga, 1995), although the value of 0.02 mgVSS/mgVSS.d (base: COD) has been cited by Lettinga et al (1996).

3.4.3.3 Net bacterial growth

The net growth is obtained by the sum of the Equations 3.12, 3.13 and 3.17:

$$\boxed{\frac{dX}{dt} = \mu.X - K_d.X} \tag{3.18}$$

or

$$\boxed{\frac{dX}{dt} = \mu_{max}.\frac{S}{K_s + S}.X - K_d.X} \tag{3.19}$$

3.4.4 Production of biological solids

3.4.4.1 Gross solids production

Bacterial growth, that is, biomass production, can be also expressed as a function of the substrate used. The greater the substrate assimilation, the greater the bacterial growth rate. This relation can be expressed as:

Growth rate = Y (Substrate removal rate)

or

$$\frac{dX}{dt} = Y\frac{dS}{dt} \tag{3.20}$$

where:
  X = concentration of microorganisms, SS or VSS (g/m^3)
  Y = yield coefficient, or coefficient of biomass production; biomass (SS or VSS) produced per unit mass of substrate removed (BOD or COD) (g/g)
  S = concentration of BOD$_5$ or COD in the reactor (g/m^3)
  t = time (d)

Therefore, it can be observed that there is a linear relationship between the bacterial growth rate and the substrate utilisation rate, or the rate of BOD or COD removal.

The value of Y can be obtained in laboratory tests with the wastewater to be treated. For the biological treatment of domestic sewage, the Y value for the heterotrophic bacteria responsible for the removal of the carbonaceous matter varies between:

- *Aerobic treatment:*

> $Y = 0.4$ to 0.8 g VSS/g BOD_5 removed (Metcalf & Eddy, 1991)
> or
> $Y = 0.3$ to 0.7 g VSS/g COD removed (EPA, 1993; Orhon and Artan, 1994)

- *Anaerobic treatment:*

> $Y \approx 0.15$ gVSS/gCOD (*acidogenic bacteria*) (van Haandel and Lettinga, 1994)
> $Y \approx 0.03$ gVSS/gCOD (*methanogenic archaea*) (van Haandel and Lettinga, 1994)
> $Y \approx 0.18$ gVSS/gCOD (*combined biomass*) (Chernicharo, 1997)

Other systems can have different Y values. The anaerobic conversion of the organic substrate releases less energy and therefore the value of Y is lower, indicating a lower biomass production. The nitrifying bacteria (chemoautotrophs) do not extract their energy from the organic carbon, but from the oxidation of inorganic compounds. Thus, they also present lower Y values when compared with the aerobic heterotrophic organisms (Arceivala, 1981).

3.4.4.2 Net solids production

Equation 3.20 expresses the gross bacterial growth without taking into consideration the reduction of the biomass due to endogenous respiration. When including the endogenous respiration, the net solids production becomes:

$$\frac{dX}{dt} = Y\frac{dS}{dt} - K_d.X \tag{3.21}$$

Example 3.2

Calculate the biological solids production in a treatment system, assuming steady state. Data:

- Reactor volume: $V = 9{,}000$ m^3
- Hydraulic detention time: $t = 3$ d
- Influent substrate (total BOD_5): $S_0 = 350$ mg/L
- Effluent substrate (soluble BOD_5): $S = 9.1$ mg/L
- Biomass in the reactor (VSS): $X_v = 173.3$ mg/L

Example 3.2 (Continued)

Coefficients of the model:

- Yield coefficient: $Y = 0.6$ mgVSS/mg BOD_5
- Endogenous respiration coefficient: $K_d = 0.06$ d^{-1}

Solution:

Assuming finite time conditions within the steady-state hypothesis, Equation 3.21 can be rewritten as:

$$\frac{\Delta X_v}{\Delta t} = 0.6\frac{gVSS}{gBOD_5}.(350-9.1)\frac{gBOD_5}{m^3}.\frac{1}{3.0d} - 0.06\frac{gVSS}{gVSS.d}.173.3\frac{gVSS}{m^3}$$

$$\Delta X_v/\Delta t = 68.2 \text{ g/m}^3.d - 10.4 \text{ g/m}^3.d = 57.8 \text{ g/m}^3.d = 0.058 \text{ kg/m}^3.d$$

Since the reactor volume is 9,000 m^3, the global net production is:

$0.058 \text{ kg/m}^3.d \times 9,000 \text{ m}^3 = 522 \text{ kgVSS/d}$

Therefore, the net production of biological solids in the system (expressed as VSS) as a function of the substrate utilisation is 522 kgVSS per day. In the calculations above, it can be seen that 68.2 g/m^3.d is the gross production and 10.4 g/m^3.d is the destruction by endogenous respiration. In this example, the net production is approximately 85% of the gross production.

In the example, numbers with decimals have been used only to clarify the calculations. In most practical applications, round figures are more frequently used for representing BOD and other variables.

3.4.4.3 Substrate removal rate

In a wastewater treatment system, it is also important to quantify the rate at which the substrate is removed. The greater the rate, the lower is the required volume for the reactor (for a certain concentration of the substrate) or the greater is the efficiency of the process (for a certain volume of the reactor).

Rearranging Equation 3.20, the substrate removal rate can be expressed as:

$$\frac{dS}{dt} = \frac{1}{Y}.\frac{dX}{dt} \qquad (3.22)$$

The substrate removal is associated with the gross biomass growth. According with Equation 3.12, $dX/dt = \mu.X$. Substituting dX/dt for $\mu.X$ in Equation 3.22, there is:

$$\boxed{\frac{dS}{dt} = \frac{\mu}{Y}.X} \qquad (3.23)$$

or (expressing μ through Equation 3.13):

$$\frac{dS}{dt} = \mu_{max} \cdot \frac{S}{K_s + S} \cdot \frac{X}{Y} \tag{3.24}$$

3.5 MODELLING OF SUBSTRATE AND BIOMASS IN A COMPLETE-MIX REACTOR

3.5.1 Mass balance in the reactor

The interactions that occur in a continuous-flow complete-mix reactor (homogenous concentration of biomass and substrate in all the reactor volume) without recirculation can be represented schematically as in Figure 3.10.

One of the characteristics of the ideal complete-mix reactor is that the effluent leaves with the same concentration as in the liquid in any part of the reactor. This implies that the values of S and X are the same in the reactor, as well as in the effluent.

X is the concentration of the solids. In the reactor, these solids are mainly biological solids, represented by the biomass (microorganisms) produced in the reactor at the expense of the available substrate. In contrast, in the influent to the reactor, the solids are those present in the wastewater, and the presence of biological solids is frequently neglected in the general mass balance. For simplicity, it is usually considered that $X_0 = 0$ mg/L (although this assumption does not apply in all situations).

Two mass balances can be done, one for the substrate and the other for the biomass. These mass balances are essential for design and operational control of the biological reactor, and are detailed in this section.

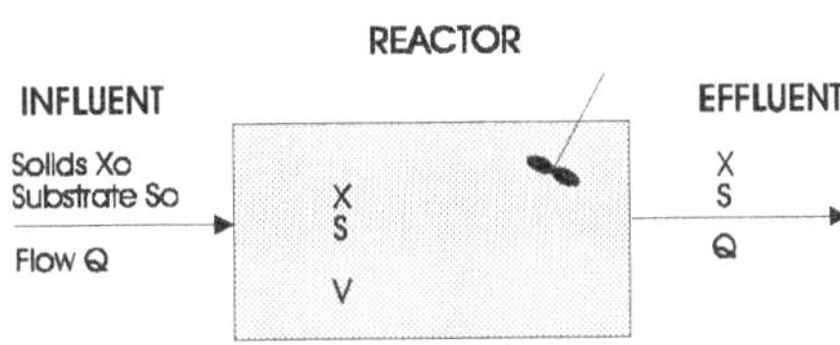

S_o	=	concentration of the total influent substrate (BOD or COD) (mg/L or g/m³)
S	=	concentration of the soluble effluent substrate (BOD or COD) (mg/L or g/m³)
Q	=	flow (m³/d)
X	=	concentration of the suspended solids in the reactor (mg/L or g/m³)
X_o	=	concentration of influent suspended solids (mg/L or g/m³)
V	=	reactor volume (m³)

Figure 3.10. Schematic representation of the mass balance in a complete-mix reactor (without recirculation)

The mass balance takes into consideration the *transport* (input and output) and the *reaction* (production and consumption) terms. The following equations are for a system composed by a single reactor, *without final settling and recirculation.*

Accumulation = Input − Output + Production − Consumption

- **Substrate balance:**

$$\frac{dS}{dt} = \frac{Q}{V}.S_o - \frac{Q}{V}.S + 0 - \frac{\mu}{Y}.X \tag{3.25}$$

where:

$$\mu = \mu_{max}.\frac{S}{K_s + S} \tag{3.26}$$

or:

$$\frac{dS}{dt} = \frac{Q}{V}.S_o - \frac{Q}{V}.S + 0 - \mu_{max}.\frac{S}{K_s + S}.\frac{X}{Y} \tag{3.27}$$

- **Solids balance:**

$$\frac{dX}{dt} = \frac{Q}{V}.X_o - \frac{Q}{V}.X + \mu.X - K_d.X \tag{3.28}$$

or:

$$\frac{dX}{dt} = \frac{Q}{V}.X_o - \frac{Q}{V}.X + \mu_{max}.\frac{S}{K_s + S}.X - K_d.X \tag{3.29}$$

3.5.2 Systems with and without solids recirculation

3.5.2.1 Introduction

There are three possible combinations of reactors with dispersed-growth biomass, continuous flow and complete-mix hydraulic regime:

- Reactor without a final sedimentation unit and hence without recirculation of solids
- Reactor with a final sedimentation unit and without recirculation of solids
- Reactor with a final sedimentation unit and with recirculation of solids

The system composed of a reactor without a final sedimentation unit and without recirculation of solids was seen in Section 3.5.1. The other systems are covered in the present section.

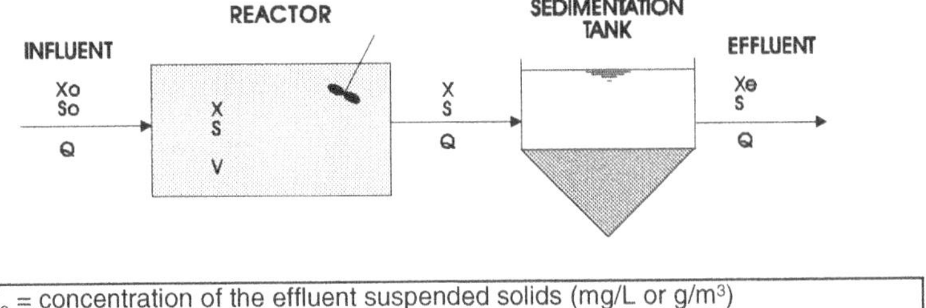

Figure 3.11.　Biological reactor followed by a settling unit (without sludge recirculation)

3.5.2.2 Reactor with a final sedimentation unit and without solids recirculation

When analysing Figure 3.10, it can be observed that the biological solids formed are present in the effluent in the same concentration as in the reactor. These solids are, ultimately, largely composed by organic matter and, if discharged into the receiving body, would undergo stabilisation similarly to the other forms of organic matter. Therefore, even that the soluble BOD may have undergone a substantial reduction in the reactor, the particulate BOD represented by the biological solids in the effluent can be responsible for the deterioration in the quantity of the effluent.

Based on this concept, various treatment systems incorporate a settling unit after the reactor in order to retain the biological solids and avoid that they reach the receiving body in the same concentration as found in the reactor. A system with a settling unit is shown in Figure 3.11.

The inclusion of a final settling unit results in a great improvement in the final effluent quality, thanks to the tendency presented by the bacteria responsible for the stabilisation of the organic matter to flocculate and settle. Thus, they have not just the property of removing BOD with efficiency, but they can be also removed by simple solid-liquid separation operations, such as sedimentation.

The capacity of a system in the removal of organic matter depends on the quantity of biomass present in the reactor. In the above system, the biomass concentration is limited by the quantity of substrate available in the influent: if the substrate increases, the bacteria population growth rate will increase, according to Monod kinetics, until a maximum limit given by μ_{max}. Hence, for a given substrate, the biomass concentration does not go above a certain maximum value.

3.5.2.3 Reactor with a final sedimentation unit and with solids recirculation

The sludge accumulated up to a certain period at the bottom of the settling unit consists mainly of bacteria that are still active in terms of their capacity to assimilate organic matter. Therefore, it is an attractive idea to use these bacteria to assist in

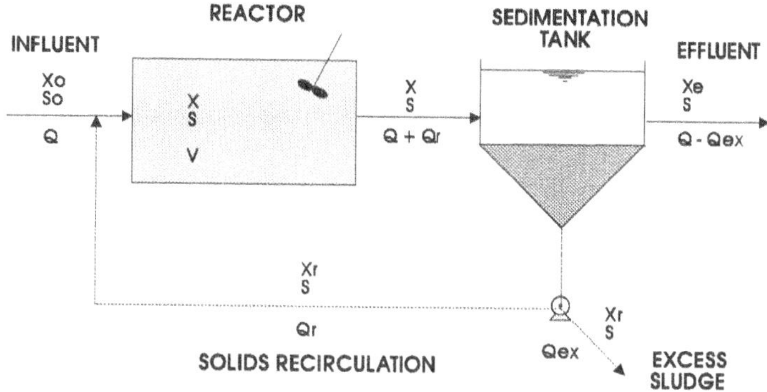

Figure 3.12.   Biological reactor with recirculation of solids

the removal of the organic matter, based on the fact that, the greater the biomass concentration, the greater the substrate utilisation or, in other words, the greater the BOD removal. Therefore, if the settled sludge is returned, with a concentration higher than in the reactor, the system will be able to assimilate a much higher BOD load. This recirculation has also the important role of increasing the average time in which the microorganisms remain in the system. The recirculation of biomass is the basic principle of systems, such as activated sludge (accomplished by a recirculation pumping station) and UASB reactors (reached by the return of solids that settled in the sedimentation tank, situated above the digestion compartment) Figure 3.12 illustrates the concept of a system with sludge return.

The value of X_r is higher than X, that is, the return sludge has a greater suspended solids concentration, what allows the increase of SS concentration in the reactor.

In Figure 3.12 there is another flow line, which corresponds to the excess sludge (also called surplus sludge, biological sludge or waste sludge). This is based on the concept that the biomass production (bacterial growth) must be compensated for by the wastage of an equivalent quantity, for the system to be maintained in equilibrium. If there were no such a wastage, the mass of suspended solids in the reactor would progressively increase, and these solids would then be transferred to the settling tank, until a point when the settler would become overloaded. In this situation, the settling tank would not be capable of transferring the solids to the bottom, and the sludge blanket level would rise and eventually the solids would start to escape in the final effluent, thus deteriorating its quality. Therefore, in a simplified way, it can be said that the production of solids must be counterbalanced by an equivalent wastage of solids (mass per unit time). The excess sludge flow is very small compared with the influent and return sludge flows.

All biological treatment systems produce excess sludge. In the complete-mix systems without recirculation (Figure 3.10), the excess sludge leaves with the final effluent. In other systems (usually with large reactor volumes), the sludge remains stored in the system and is only removed after large time intervals. The system represented in Figure 3.11 could be according to this concept (e.g. complete-mix aerated lagoons followed by sedimentation ponds) or it could include a separate line for the continuous or periodic removal of the excess sludge.

3.5.3 Hydraulic detention time and solids retention time

In a system with solids recycling, such as in Figure 3.12, the solids are separated and concentrated in the final settling unit and subsequently returned to the reactor. The liquid, on the other hand, in spite of the recirculation (which is internal in the system), does not vary quantitatively, apart from the withdrawal of the excess sludge flow, which is negligible in the overall calculation ($Q_{ex} \approx 0$). Therefore, only the solids are retained in the system, owing to the separation, thickening and recycling. Thus, the solids remain longer in the system than the liquid. It is thus necessary to distinguish the concepts of *solids retention time* and *hydraulic detention time*. Other treatment systems retain solids without the need of separate settling tanks (e.g. sequencing batch reactors) or recycle pumps (e.g. sequencing batch reactors, UASB reactors).

The **hydraulic detention time** t (or hydraulic retention time – HRT) given by:

$$\text{hydraulic detention time} = \frac{\text{volume of liquid in the system}}{\text{volume of liquid removed per unit time}} \tag{3.30}$$

Since the volume of liquid that enters is the same as the one that leaves, the following generalisation can be made:

$$\boxed{t = \frac{V}{Q}} \tag{3.31}$$

Similarly, the **solids retention time** SRT (or *mean cell residence time* – MCRT or *sludge age* - θ_c) is given by:

$$\text{sludge age} = \frac{\text{mass of solids in the system}}{\text{mass of solids produced per unit time}} \tag{3.32}$$

In the steady state, the quantity of solids removed from the system is equal to the quantity of sludge produced. Hence, the sludge age can also be expressed as:

$$\text{sludge age} = \frac{\text{mass of solids in the system}}{\text{mass of solids removed per unit time}} \tag{3.33}$$

Since the biomass production can be represented by dX/dt, Equation 3.33 can be written as:

$$\theta_c = \frac{V.X}{V.\left(\dfrac{dX}{dt}\right)} = \frac{X}{dX/dt} \qquad (3.34)$$

As seen in Section 3.4.3.3, the net bacterial growth (denominator of Equation 3.34) can be given by:

$$\frac{dX}{dt} = \mu.X - K_d.X = (\mu - K_d).X \qquad (3.35)$$

Substituting Equation 3.35 into 3.34, the equation that represents the sludge age is obtained:

$$\theta_c = \frac{1}{\mu - K_d} \qquad (3.36)$$

Depending on inclusion or not of sludge recycle, the following two conditions are obtained:

- Systems without solids retention: $t = \theta_c$
- Systems with solids retention: $t > \theta_c$

The fact that the biomass stays longer than the liquid in the system justifies the greater efficiency of systems with solids recirculation, compared with systems without solids recirculation. It can also be said that, for the same removal efficiency, systems with solids recirculation require much smaller reactor volumes than the systems without recirculation.

In all the above analyses, the following simplifying hypotheses have been adopted:

- *The biochemical reactions occur only in the reactor.* The reactions of the conversion of organic matter and of cellular growth in the settling unit can be neglected, when compared with those that occur in the reactor. The error resulting from this simplification can be considered negligible.
- *The biomass is assumed to be present only in the reactor.* In the calculation of the sludge age, the solids present in the final settling unit and in the recirculation line have not been considered. This is only a question of convention, and normally only the mass in the reactor is considered, due to the greater simplicity associated with this procedure (measurement of only the SS concentration in the reactor). If the component of the mass of

solids present in the settling tank is incorporated, this needs to be clearly stated when presenting the sludge age value.

- *The mechanisms take place according to the steady state.* This hypothesis greatly simplifies the real situation that occurs in a wastewater treatment plant, in which the true steady state will practically never occur. The continuous variation of the influent characteristics (throughout the day) is responsible for the predominance of the dynamic state in operation, with mass accumulations occurring in the reactor and settling tank. However, if the system is analysed in a broad time scale, these variations become less important. Thus, it can be said that, for designing or operational planning in long time horizons, the steady state assumption can be accepted. On the other hand, for the operation of a plant in a short time scale, the predominance of the dynamic state must be taken into consideration, and the above relations cannot be used as such. *In the dynamic state, the mass of solids produced is not equal to the mass wasted, which alters the interpretation of the sludge age concept.*
- *The influence of the solids in the influent sewage was not considered.* This is a simplifying assumption adopted in most books, but it can be far from reality in some sewage treatment plants with a lower production of biological solids.

Example 3.3

Calculate the hydraulic detention time and the sludge age in the sewage treatment system described in Example 3.2 (without a settling tank and solids recirculation). The main relevant data from Example 3.2 are:

Reactor volume: $V = 9{,}000 \text{ m}^3$
Input and output variables:

- Influent flow: $Q = 3{,}000 \text{ m}^3/\text{d}$
- Influent substrate (BOD$_5$ total): $S_o = 350 \text{ mg/L}$
- Effluent substrate (BOD$_5$ soluble): $S = 9.1 \text{ mg/L}$

Model coefficients:

- Maximum specific growth rate: $\mu_{max} = 3.0 \text{ d}^{-1}$
- Half-saturation coefficient: $K_s = 60 \text{ mg/L}$
- Endogenous respiration coefficient: $K_d = 0.06 \text{ d}^{-1}$

Solution:

a) Hydraulic detention time
    From Equation 3.31:

$$t = \frac{V}{Q} = \frac{9{,}000 \text{ m}^3}{3{,}000 \text{ m}^3/\text{d}} = 3.0 \text{ d}$$

Example 3.3 (Continued)

b) Sludge age
The value of μ is given by Equation 3.13:

$$\mu = \mu_{max}\cdot\frac{S}{K_s + S} = 3.0.\frac{9.1}{60 + 9.1} = 0.395\,\mathrm{d^{-1}}$$

The sludge age is given by Equation 3.36:

$$\theta_c = \frac{1}{\mu - K_d} = \frac{1}{0.395 - 0.06} = 3.0\,\mathrm{d}$$

As expected, in the present example $t = \theta_c$, since the system has no solids recirculation.

3.5.4 Cell wash-out time

The time in which the bacteria stays in the treatment system (θ_c) must be higher than the time necessary for its duplication. If not, the cell is going to be washed out of the system before it has time to multiply itself, leading to a progressive reduction in the biomass concentration in the reactor, until the system collapses.

As previously seen, the reproduction of bacteria is by binary fission, and the net specific growth rate is as seen in Section 3.4.3:

$$\frac{dX}{dt} = (\mu - K_d).X \tag{3.37}$$

or

$$\frac{dX}{X} = (\mu - K_d).dt \tag{3.38}$$

The integration of this equation within the limits of $t = 0$ and $t = t$ leads to:

$$\ln\frac{X}{X_o} = (\mu - K_d).t \tag{3.39}$$

where:
  $X =$ number or bacterial concentration at a time t
  $X_0 =$ number or bacterial concentration at a time $t = 0$

This is the exponential growth phase, which, if plotted on a logarithmic scale, gives a straight line. The duplication time is that in which $X = 2X_0$, or:

$$\ln 2 = (\mu - K_d).t \tag{3.40}$$

Hence, the doubling time t_{dup} is given by:

$$t_{dup} = \frac{\ln 2}{\mu - K_d} = \frac{0.693}{\mu - K_d} \tag{3.41}$$

The considerations here are also distinct for the systems with and without recirculation (Arceivala, 1981):

- *Systems with suspended biomass, without recirculation* ($\theta_c = t$). In this case, θ_c ($= t$) needs to be greater than or equal to t_{dup}. This condition needs to be satisfied in the case, for instance, of complete-mix aerated lagoons. In these lagoons, it is essential to ensure that the minimum hydraulic detention time is not less than the bacterial doubling time, at critical temperature conditions. In the design of units such as ponds in series, the minimum size of each pond is dictated by this requirement.
- *Systems with suspended biomass, with recirculation* ($\theta_c > t$). In these systems, the excess sludge flow can be adjusted in order to maintain $\theta_c > t_{dup}$, while the hydraulic detention time t can be maintained at a minimum (minimum volume of the reactor). Consequently, the sludge recirculation is a way of increasing θ_c without necessarily increasing t (or V).

For the aerobic removal of the carbonaceous matter, the solids retention time of the heterotrophic bacteria is usually much higher than the minimum time required. However, for the methanogenesis in anaerobic systems, as well as for the oxidation of ammonia in aerobic systems, greater care must be exercised. The reproduction rate of the methanogenic and nitrifying organisms is very slow and there is the risk of their wash-out of the system if the influent flow increases substantially or if their reproduction rate is reduced due to some environmental problem.

3.5.5 Concentration of suspended solids in the reactor

To obtain the biomass concentration in the reactor of a system **with solids recirculation**, equations 3.21 and 3.34 can be rearranged, assuming steady-state conditions:

$$X = \frac{Y.(S_o - S)}{1 + K_d.\theta_c} \cdot \left(\frac{\theta_c}{t}\right) \tag{3.42}$$

This equation is very important for the estimation of the solids concentration in a complete-mix reactor, once the other parameters or variables are known or have been estimated. The analysis of this equation also leads to interesting considerations about the influence of the sludge recirculation on the SS concentration in the reactor.

In a system **without recirculation** it was seen that $\theta_c = t$. Consequently, Equation 3.42 is reduced to:

$$\boxed{X = \frac{Y.(S_0 - S)}{1 + K_d.t}}$$ (3.43)

It can be observed that the difference between both equations is the factor (θ_c/t), which exerts a multiplying factor on Equation 3.43, in the sense of increasing the suspended solids concentration in the reactor. In the design of a wastewater treatment plant, any increase in X allows a proportional decrease in the volume required for the reactor. Example 3.4 illustrates the calculation of the biomass concentration in a system without recirculation, while Example 3.5 shows the advantage of the recirculation in terms of the reduction of the reactor volume.

Example 3.4

Calculate the suspended solids concentration to be reached, under steady-state conditions, in the reactor described in Example 3.2. The relevant data from Example 3.2 are:

- Influent substrate (total BOD_5): $S_0 = 350$ mg/L
- Effluent substrate (soluble BOD_5): $S = 9.1$ mg/L
- Hydraulic detention time: $t = 3.0$ days (obtained in Example 3.3)
- Yield coefficient: $Y = 0.6$ mgVSS/mgBOD$_5$
- Endogenous coefficient: $K_d = 0.06$ d^{-1}

Solution:

From Equation 3.43:

$$X_v = \frac{Y.(S_0 - S)}{1 + K_d.t} = \frac{0.6 \times (350 - 9.1)\,\text{mg/L}}{1 + 0.06\text{d}^{-1} \times 3\text{d}} = 173.3\ \text{mg/L}$$

This value was used as input data in Example 3.2.

Example 3.5

Calculate the biomass concentration in the reactor, for the following conditions:

- system without recirculation, $t = \theta_c = 5$ days (e.g.: complete-mix aerated lagoon)
- system with recirculation, $t = 0.25$ days (6 hours) and $\theta_c = 5$ days (e.g.: conventional activated sludge)

Example 3.5 (Continued)

Adopt $Y = 0.6$; $K_d = 0.07\,d^{-1}$; $S_o = 300\,mg/L$; $S = 15\,mg/L$

Solution:

a)   System without recirculation

From Equation 3.43:

$$X_v = \frac{Y.(S_o - S)}{1 + K_d.t} = \frac{0.6 \times (300 - 15)\,mg/L}{1 + 0.07d^{-1} \times 5d} = 127\,mg/L$$

b)   System with recirculation

From Equation 3.42:

$$X_v = \frac{Y.(S_o - S)}{1 + K_d.\theta_c} \cdot \left(\frac{\theta_c}{t}\right) = \frac{0.6 \times (300 - 15)\,mg/L}{1 + 0.07d^{-1} \times 5d} \cdot \left(\frac{5\,d}{0.25\,d}\right) = 2.540\,mg/L$$

c)   Comments

It can be observed that in the system with sludge recirculation, the VSS concentration 2,540 mg/L is much higher than in the option without recirculation, in which VSS is equal to 127 mg/L. The ratio between the two concentrations is:

    2540 mg/L/127 mg/L $= 20$

This is the same ratio between the hydraulic detention times in the reactors of the two systems:

    5 d / 0.25 d $= 20$

In other words, it can be said in this example that, for the same effluent characteristics, the volume of the reactor in the system with recirculation is 20 times less than in the system without recirculation.

3.5.6 Effluent substrate

According with Equation 3.36, the sludge age can be expressed as:

$$\boxed{\theta_c = \frac{1}{\mu - K_d}} \tag{3.44}$$

The rearrangement of Equation 3.44 leads to:

$$\frac{1}{\theta_c} = \mu - K_d \tag{3.45}$$

or:

$$\frac{1}{\theta_c} = \mu_{max} \cdot \left(\frac{S}{K_s - S}\right) - K_d \tag{3.46}$$

Rearranging Equation 3.46 in terms of S:

$$S = \frac{K_s \cdot [(1/\theta_c) + K_d]}{\mu_{max} - [(1/\theta_c) + K_d]} \tag{3.47}$$

This is the general equation to estimate the soluble effluent BOD from a complete-mix reactor. An interesting aspect of this equation is that, mathematically, in a complete-mix system, the effluent BOD concentration S is independent of the influent concentration S_o (Arceivala, 1981). This is because K_s, K_d and μ_{max} are constants and therefore S depends only on the sludge age θ_c. This can be understood by the fact that, the greater the influent BOD, the greater is the production of biological solids, and, as a result, the greater is the biomass concentration X_v. Thus, when there is more food, there is a greater availability for the bacteria to assimilate it. It must be emphasised that this consideration is applicable only in the steady state. In the dynamic state, the increases in the influent BOD are not immediately followed by the corresponding increase in the biomass, since the process of biomass increase is slow. Hence, until a new equilibrium state is reached (if it will be reached at all), the effluent quality, in terms of BOD, will be deteriorated.

Theoretically, the minimum concentration of soluble substrate that can be reached in a system is when the sludge age θ_c tends to infinity. In these conditions, the term $1/\theta_c$ is equal to zero. By substituting $1/\theta_c$ for 0 in Equation 3.47, an equation that defines the minimum possible soluble effluent BOD (S_{min}) is found. If in a treatment system it is necessary to obtain a value lower than S_{min}, it will not be possible with only one complete-mix reactor (Grady & Lim, 1980). S_{min} is independent on the existence of recirculation and is only a function of the kinetic coefficients.

$$S_{min} = \frac{K_s \cdot K_d}{\mu_{max} - K_d} \tag{3.48}$$

Example 3.6

Calculate the soluble effluent BOD concentration, after steady-state conditions have been reached in the system described in Example 3.2. Since the system has no solids recirculation, the sludge age is equal to the hydraulic detention time. The relevant data for this example are:

- Sludge age (= equal to the hydraulic detention time): $\theta_c = t = 3.0$ d (according with Example 3.3)

Example 3.6 (Continued)

- Maximum specific growth rate: $\mu_{max} = 3.0 \text{ d}^{-1}$
- Half-saturation coefficient: $K_s = 60 \text{ mg/L}$
- Endogenous respiration coefficient: $K_d = 0.06 \text{ d}^{-1}$

Solution:

From Equation 3.47:

$$S = \frac{K_s.[(1/\theta_c) + K_d]}{\mu_{max} - [(1/\theta_c) + K_d]} = \frac{60 \times [(1/3) + 0.06]}{3.0 - [(1/3) + 0.06]} = 9.1 \text{mg/L}$$

This value of S = 9.1 mg/L is the same as the one adopted in the preceding examples.

3.5.7 Loading rates on biological reactors

3.5.7.1 Sludge load (food-to-microorganism ratio)

A relationship widely used by designers and operators of wastewater treatment plants is the *sludge load* or *F/M* (food-to-microorganism) *ratio*. It is based on the concept that the quantity of food or substrate available per unit mass of microorganisms is related to the efficiency of the system. Hence, it can be understood that, the higher the BOD load supplied per unit value of the biomass (high F/M ratio), the lower is the substrate assimilation efficiency, but, on the other hand, the lower is the required reactor volume. Conversely, when less BOD is supplied to the bacteria (low F/M ratio), the demand for food is higher, which implies a greater BOD removal efficiency and a larger reactor volume requirement. In a situation in which the quantity of food supplied is very low, the mechanism of endogenous respiration becomes prevalent, which is a characteristic of low-rate (e.g. extended aeration) systems.

The food load supplied is given by:

$$F = Q.S_0 \tag{3.49}$$

The microorganism mass is calculated as:

$$M = V.X_v \tag{3.50}$$

where:

Q = influent flow (m^3/d)
S_0 = influent BOD$_5$ concentration (g/m^3)
V = reactor volume (m^3)
X_v = volatile suspended solids concentration (g/m^3)

Thus, the F/M ratio is expressed as:

$$\frac{F}{M} = \frac{Q.S_0}{V.X_v}$$ (3.51)

where:

F/M = sludge load (gBOD$_5$ supplied per day/g VSS)

The F/M ratio is sometimes expressed in terms of total suspended solids (TSS) instead of VSS. Care must be exercised when analysing values, in order not to mix nomenclatures and forms of expression. There is a relationship between VSS and SS, which is a function of the sludge age. High sludge ages (low F/M ratios) imply higher removals of the organic fraction, represented by the volatile suspended solids, leading to a lower VSS/SS ratio. Each wastewater treatment system has typical values of the VSS/SS ratio predominant in the biological reactor.

The relation Q/V in Equation 3.51 can be substituted by 1/t, which leads to another way of presenting the F/M ratio:

$$\frac{F}{M} = \frac{S_0}{t.X_v}$$ (3.52)

Accurately speaking, the F/M ratio has no direct association with the removal of the organic matter that really occurs in the reactor, since the F/M ratio constitutes only a representation of the *applied* (or available) load. The formula that expresses the relation between the available and the removed substrates is the *substrate utilisation rate* (U). In U, instead of including only S_0, the relation $S_0 - S$ is included:

$$U = \frac{Q.(S_0 - S)}{V.X_v}$$ (3.53)

where:

S = soluble effluent BOD$_5$ concentration (g/m^3)

Hence, it can be said that:

$$U = (F/M).E$$ (3.54)

where:

E = substrate removal efficiency of the system = $(S_0 - S)/S_0$

Because the substrate removal efficiencies in sewage treatment systems are usually high and not far from the unity, it can be said that $U \approx F/M$.

Analysing Equation 3.53, it can be seen that, after defining the design value for F/M (or U), adopting a value for X_v, and knowing the flow Q and the influent S_o and effluent S (desired) BOD concentrations, the necessary volume for the reactor can be calculated. Rearranging the formula leads to:

$$V = \frac{Q.(S_o - S)}{X_v.U} \qquad (3.55)$$

Example 3.7 illustrates the use of the concepts of F/M and U to analyse the operational range of an existing works, while Example 3.8 presents U as a design parameter in the calculation of the reactor volume.

Example 3.7

Calculate the values of F/M and U in a wastewater treatment plant with sludge recirculation, as described in Example 3.5. Data:

$S_o = 300 \text{ gBOD}_5/\text{m}^3$
$S = 15 \text{ gBOD}_5/\text{m}^3$
$t = 0.25 \text{ d}$
$X_v = 2{,}540 \text{ gVSS/m}^3$

Solution:

a) Calculation of F/M

From Equation 3.52:

$$\frac{F}{M} = \frac{S_o}{t.X_v} = \frac{300 \text{ gBOD}_5/\text{m}^3}{0.25 \text{ d} . 2{,}540 \text{ gVSS/m}^3} = 0.47\text{d}^{-1}$$

F/M = 0.47 kgBOD$_5$/kgVSS.d

b) Calculation of U

From Equation 3.53, substituting Q/V by 1/t:

$$\frac{F}{M} = \frac{S_o - S}{t.X_v} = \frac{(300 - 15)\,\text{gBOD}_5/\text{m}^3}{0.25 \text{ d} . 2{,}540 \text{ gVSS/m}^3} = 0.45\text{d}^{-1}$$

U = 0.45 kgBOD$_5$/kgVSS.d

In comparison with Example 3.5, is can be seen that, in this case, the substrate utilisation rate U is equal to 0.45 kgBOD$_5$/kgVSS.d, corresponding to a sludge age of 5.0 days (calculated in Example 3.5).

Example 3.8

Calculate the reactor volume of an extended aeration activated sludge system, given that:

$U = 0.12\ kgBOD_5/kgVSS.d$ (adopted)
$Q = 5,000\ m^3/d$ (design data)
$S_o = 340\ mg/L$ (design data)
$S = 5\ mg/L$ (desired)
$X_v = 3,500\ mg/L$ (adopted)

Solution:

From Equation 3.55:

$$V = \frac{Q.(S_o - S)}{X_v.U} = \frac{5,000\ m^3/d.(340 - 5)gBOD_5/m^3}{3,500\ gVSS/m^3.0.12\ kgBOD_5/kgVSS.d} = 3,988\ m^3$$

3.5.7.2 Relationship between the substrate utilisation rate (U) and the sludge age (θ_c)

In the steady state, there is no accumulation of solids in the system, which makes the following relation valid:

Solids production rate $=$ *Solids removal rate*
(biological solids generated) $=$ *(excess sludge wasted)*

From Equation 3.21:

$$\boxed{\frac{dX}{dt} = Y.\frac{dS}{dt} - K_d.X} \tag{3.56}$$

or:

$$\boxed{\frac{\Delta X_v}{\Delta t} = Y.\frac{(S_o - S)}{t} - K_d.f_b.X_v} \tag{3.57}$$

Dividing by X_v:

$$\frac{\Delta X_v/\Delta t}{X_v} = Y.\frac{(S_o - S)}{X_v.t} - K_d.f_b \tag{3.58}$$

According to Equation 3.34, the first part of Equation 3.58 is equal to $1/\theta_c$. Hence:

$$\frac{1}{\theta_c} = Y.\left(\frac{S_o - S}{X_v.t}\right) - K_d.f_b \qquad (3.59)$$

According to Equation 3.53, the term in brackets is the substrate utilisation rate (U). Therefore, in the steady state:

$$\frac{1}{\theta_c} = Y.U - K_d.f_b \qquad (3.60)$$

Since $U = (F/M).E$, Equation 3.60 can also be also expressed as:

$$\frac{1}{\theta_c} = Y.\left(\frac{F}{M}\right).E - K_d.f_b \qquad (3.61)$$

Therefore, Equations 3.60 and 3.61 correlate θ_c and U (or F/M), once the values of the coefficients Y and K_d have been adopted. Thus, knowing θ_c, U can be calculated (or F/M), or vice-versa.

Equations 3.60 and 3.61 are presented in various texts in terms of X_V, without considering that only part of X_V is biodegradable. Hence, the correct is to express the decay in terms of X_b, or of $f_b.X_v$, since $f_b = X_b/X_v$, where X_b is the concentration of the biodegradable volatile suspended solids and f_b is the biodegradable fraction of VSS.

The reason for the consideration of X_b is reinforced by the following point. In systems with a high sludge age, such as the extended aeration activated sludge, if Equations 3.60 and 3.61 were presented without f_b, certain indeterminations could be generated (not present in conventional texts, because most of them are dedicated to systems with a conventional sludge age). When substituting Y and K_d by typical values in the literature and adopting F/M values representative of extended aeration, a negative value of $1/\theta_c$ (negative net sludge production) is obtained. Such incongruence occurs because the values of Y and K_d expressed in the literature are mainly associated to determinations undertaken in systems with a conventional sludge age. A solution for this problem is the adoption of coefficient values resulting from laboratory tests under operational conditions similar to those expected in practice. The main discrepancy found in long sludge age systems is because of the fact that, given the low substrate availability prevalent in the medium, a predominance of endogenous respiration occurs, which causes a decrease of the biodegradable fraction of the volatile solids. Consequently, the higher the sludge age, the larger is the inert fraction of the solids (due to the aerobic digestion that takes place in the reactor), which results in a lower biodegradable fraction f_b. The concept of f_b is detailed in Section 3.5.8.

The selection of the K_d value should also reflect the representation of the biomass. The values of K_d usually reported in the literature are associated with the decay of X_v. In this text, when representing the decay in terms of the biodegradable solids, K_d assumes higher values than usually found in the literature, because it will be subsequently multiplied by f_b. When multiplying the proposed values of f_b and K_d, in a system with a conventional sludge age, the results obtained correspond to the usual values of K_d, found in the literature.

The present approach, described by Eckenfelder (1989), allows the generalisation of systems such as activated sludge for the conventional as well as for the extended aeration mode. There are even more sophisticated models, such as those of the IWA (2000) – ASM1-1987, ASM2-1995, ASM3-2000 – based on the active fraction of the biomass. These models, in spite of their great explanatory power, scientifically based structure and wide acceptance, are more complex, being outside of the more general scope of this book. However, readers who want to deepen their knowledge into the mechanisms of C, N and P removal, are encouraged to consult these references, especially the latter, which is a consolidation of the previous two.

3.5.7.3 Volumetric organic load

Some sewage treatment systems are designed according to the *volumetric organic load*, which is represented by the equation:

$$L_{VO} = \frac{Q.S_o}{V} \tag{3.62}$$

where:

L_{VO} = volumetric organic load (gBOD/m^3.d)

Since Q/V is equal to $1/t$, Equation 3.62 can be rewritten as:

$$L_{VO} = \frac{S_o}{t} \tag{3.63}$$

Adopting a value for L_{VO}, the required reactor volume can be calculated through the rearrangement of Equation 3.62:

$$V = \frac{Q.S_o}{L_{VO}} \tag{3.64}$$

It can be observed that the volumetric organic load differs from the F/M ratio by the fact that the former represents the load applied per unit reactor volume, while the latter represents the load applied per unit of biomass in the reactor.

Example 3.9

Calculate the volumetric organic load in the wastewater treatment system with sludge recirculation, as described in Example 3.5. Data:

$S_o = 300 \text{ gBOD}_5/\text{m}^3$
$t = 0.25 \text{ d}$

Solution:

From Equation 3.63:

$$L_{VO} = \frac{S_o}{t} = \frac{300 \text{gDBO}_5/\text{m}^3}{0.25\text{d}} = 1,200 \text{g}/\text{m}^3.\text{d}$$

3.5.7.4 Volumetric hydraulic load

Other wastewater treatment systems are designed based on the *volumetric hydraulic load* (L_V):

$$L_V = \frac{Q}{V} \tag{3.65}$$

where:
L_V = volumetric hydraulic load (m^3/d per m^3)

It should be observed that L_V is simply the reciprocal of the hydraulic detention time t ($L_V = 1/t$).

3.5.7.5 Surface organic load

Still other biological reactors, such as facultative ponds, are designed based on a *surface organic load*, or *surface loading rate* (L_s). In these reactors, the surface area plays a more important role than the volume itself. In the case of facultative ponds, it is through the surface area that the sunlight penetrates in the pond and allows the development of the required photosynthetic activity. Surface loading rates are expressed as:

$$L_S = \frac{Q.S_o}{A} \tag{3.66}$$

where:
L_S = surface organic load ($\text{kgBOD}/\text{m}^2.\text{d}$ or kgBOD/ha.d)
A = surface area (m^2 or ha)

3.5.8 Distribution of the biological solids in the treatment

As already seen, the total suspended solids are composed by an *inorganic (fixed) fraction* (X_i) and an *organic (volatile) fraction* (X_v):

$$X = X_i + X_v$$ (3.67)

On the other hand, another division should still be established, because not all the volatile suspended solids are biodegradable. In the volatile solids, there is a fraction, which is *non-biodegradable (inert)* (X_{nb}), resulting from residues of endogenous respiration, and a *biodegradable* fraction (X_b). Thus:

$$X_v = X_{nb} + X_b$$ (3.68)

The sludge recirculation leads to an accumulation of the inorganic fraction X_i, as well as the non-biodegradable fraction X_{nb} in the system, since they are not affected by the biological treatment. The higher the sludge age, the lower the ratio X_b/X_v. This can be understood by the fact that at higher sludge ages there is the predominance of endogenous respiration with a greater self-oxidation of the cellular material, that is, stabilisation of the sludge

The volatile solids shortly after being produced ($\theta_c = 0$) are approximately 20% inert and 80% biodegradable. With their stay in the reactor ($\theta_c > 0$), the ratio X_b/X_v decreases. The ratio X_b/X_v ($= f_b$) can be expressed as (Eckenfelder, 1989):

$$f_b = \frac{f_b'}{1 + (1 - f_b').K_d.\theta_c}$$ (3.69)

where:

f_b = biodegradable fraction of the VSS generated in the system, submitted to a sludge age θ_c (X_b/X_v)

f_b' = biodegradable fraction of the VSS immediately after their generation in the system, that is, with $\theta_c = 0$. This value is typically equal to 0.8 ($= 80\%$)

For the various values of K_d and sludge age, Table 3.1 presents the f_b values resulting from Equation 3.69.

The values of f_b are used in various formulas in the activated sludge process, such as those related with the sludge production, oxygen consumption by the biomass and BOD associated with the suspended solids in the effluent.

The values presented in Table 3.1 are only related to the biological solids produced in the reactor. Raw sewage also contributes with fixed solids and volatile,

Table 3.1. Biodegradable fraction of the VSS (f_b), according to Equation 3.69, for various values of θ_c and K_d

θ_c (days)	$X_b/X_v(= f_b)$ ratio			
	$K_d = 0.05$ d^{-1}	$K_d = 0.07$ d^{-1}	$K_d = 0.09$ d^{-1}	$K_d = 0.11$ d^{-1}
4	0.77	0.76	0.75	0.74
8	0.74	0.72	0.70	0.68
12	0.71	0.68	0.66	0.63
16	0.69	0.65	0.62	0.59
20	0.67	0.63	0.59	0.56
24	0.65	0.60	0.56	0.52
28	0.63	0.57	0.53	0.50
32	0.61	0.55	0.51	0.47

non-biodegradable and biodegradable solids. Approximate values of the main relationships in raw sewage are (WEF/ASCE, 1992; Metcalf & Eddy, 1991):

Raw sewage:

- VSS/TSS $= 0.70 - 0.85$
- SS$_i$/TSS $= 0.15 - 0.30$

- SS$_b$/VSS $= 0.6$
- SS$_{nb}$/VSS $= 0.4$

The load relative to the contribution of these raw sewage solids should be taken into account, especially the inorganic and non-biodegradable organic fractions, which do not undergo transformation in the biological treatment. The load of biodegradable solids need not be taken into separate consideration, since these solids will be absorbed in the biological flocs in the reactor, where they will be hydrolysed and subsequently degraded, thus generating new biological solids and an oxygen consumption (in aerobic systems). Since this contribution is already included in the biological solids generated due to the influent BOD, the SS$_b$ of the raw sewage should not be calculated separately. In systems with primary settling, the fraction of the raw sewage solids that is removed by sedimentation and do not enter the biological reactor should be discounted.

The *active fraction* of the volatile solids, that is, the fraction that is really responsible for the decomposition of the carbonaceous organic matter, is given by (IAWPRC, 1987; WEF/ASCE, 1992):

$$f_a = \frac{1}{1 + (1 - f_b').K_d.\theta_c} \tag{3.70}$$

where:

f_a = active fraction of the volatile suspended solids (X_a/X_v)

The fraction f_a can be also expressed as:

$$f_a = f_b/f_b'$$
(3.71)

Some mathematical models (IWA, 2000) express the kinetics of the removal of organic matter in terms of the active suspended solids. However, in the present book, the more conventional and simpler version of expressing the solids by VSS is adopted.

During the treatment, there is the generation and destruction of solids. This section presents below the various formulas related to the creation and destruction of the various types of solids present in the reactor (excluding the solids of the influent wastewater).

The **gross production of volatile suspended solids** (P_{xv}) is a result of the multiplication of the yield coefficient by the BOD_5 load removed:

$$P_{xv}\ gross = Y.Q.(S_0 - S)$$
(3.72)

From the suspended solids recently formed in the reactor, approximately 90% are organic (volatile) and 10% are inorganic (fixed) (Metcalf & Eddy, 1991), leading to the ratio VSS/TSS = 0.9 in the recently formed biological solids. With this ratio, the recently produced TSS load can be estimated (**gross production of total suspended solids**, without including destruction):

$$P_x\ gross = P_{xv}/0.9$$
(3.73)

As a result, the **production of fixed suspended solids** is:

$$P_{xi} = P_x\ gross - P_{xv}\ gross$$
(3.74)

Not all the volatile solids produced are biodegradable. Immediately after production $(\theta_c = 0)$, the load of the biodegradable solids produced is equal to the product of the volatile solids produced (P_{xv}) by the biodegradable fraction of the solids recently formed (f_b'). It was seen above that typical values of f_b' are around 0.8. The determination of this load of **recently formed biodegradable suspended solids** has little practical value, since the solids remain in the reactor for a time greater than θ_c. Only as an illustration, the formula for its calculation is presented below:

$$P_{xb}\ recently\ formed = (P_{xv}\ gross).f_b'$$
(3.75)

As a result of the time that the solids remain in the reactor (θ_c), the biodegradability fraction f_b decreases. Hence, the **gross production of biodegradable**

suspended solids submitted to a detention time θ_c is equal to P_{xv} gross multiplied by the biodegradability fraction f_b. As seen above, f_b is a function of θ_c (Equation 3.69).

$$\boxed{P_{xb}\ \text{gross} = (P_{xv}\ \text{gross}).f_b} \tag{3.76}$$

The **production of non-biodegradable volatile suspended solids** (inert or endogenous) is obtained by the difference between the gross production of X_v and the gross production of X_b:

$$\boxed{P_{xnb} = P_{xv}\ \text{gross} - P_{xb}\ \text{gross}} \tag{3.77}$$

The **gross production of the active fraction of the volatile suspended solids** is given by:

$$\boxed{P_{xa}\ \text{gross} = (P_{xv}\ \text{gross}).f_b/f_b'} \tag{3.78}$$

Owing to endogenous respiration, part of the biodegradable solids are destroyed in the reactor. The **load of the biodegradable suspended solids destroyed** is a function of the sludge age, and is given by:

$$\boxed{P_{xb}\ \text{destroyed} = (P_{xb}\ \text{gross}).(K_d.\theta_c)/(1 + f_b.K_d.\theta_c)} \tag{3.79}$$

Therefore, the **net production of biodegradable suspended solids** is:

$$\boxed{P_{xb}\ \text{net} = P_{xb}\ \text{gross} - P_{xb}\ \text{destroyed}} \tag{3.80}$$

The **net production of volatile suspended solids** is equal to the net production of biodegradable solids plus the production of non-biodegradable organic solids.

$$\boxed{P_{xv}\ \text{net} = P_{xb}\ \text{net} + P_{xnb}} \tag{3.81}$$

The net production of the volatile suspended solids can also be obtained by using the concept of the observed yield (Y_{obs}). Y_{obs} already takes into consideration the destruction of the biodegradable solids and is expressed by:

$$\boxed{Y_{obs} = \frac{Y}{1 + f_b.K_d.\theta_c}} \tag{3.82}$$

Thus, the net production of VSS can also be given by:

$$P_{xv} \text{ net} = Y_{obs}.Q.(S_o - S)$$ (3.83)

The **net production of the total suspended solids** is equal to the net production of volatile solids plus the net production of inorganic solids:

$$P_x \text{ net} = P_{xv} \text{ net} + P_{xi}$$ (3.84)

The *final ratio VSS/TSS* (due only to biological solids) in the reactor is obtained through:

$$VSS/SS = (P_{xv} \text{ net})/(P_x \text{ net})$$ (3.85)

The *percentage destruction of the biodegradable suspended solids* due to endogenous respiration is given by:

$$\% \text{ destruction } X_b = 100.(P_{xb} \text{ destroyed})/(P_{xb} \text{ gross})$$ (3.86)

The *percentage destruction of the volatile suspended solids* is expressed by:

$$\% \text{ destruction } X_v = 100.(P_{xb} \text{ destroyed})/(P_{xv} \text{ gross})$$ (3.87)

If it is desired to take into account the solids present in the influent wastewater, the load of the influent solids (inorganic solids and non-biodegradable solids) should be added to each one of the loads of biological solids produced. As mentioned above, it is not necessary to sum the contribution of the biodegradable solids of the influent wastewater, due to the fact that they are already included in the influent BOD. The influent BOD comprises the soluble BOD_5 fraction as well as the particulate BOD_5 fraction (resulting from the influent biodegradable solids). Consequently, the transformation of the total BOD_5 into biodegradable solids indirectly incorporates the contribution of the biodegradable solids from the influent wastewater.

Example 3.10 synthesises the relations described above for conventional and extended aeration activated sludge systems. In order to allow a comparison, the starting point in both cases is the removal of a BOD load equal to $100 \text{ kgBOD}_5/d$. Hence, the values of the production and destruction, as well as the distribution of the solids, are referenced to a value of 100 units. In both situations, only the biological solids formed in the reactor are analysed and the contribution of the solids from the influent wastewater are not taken into consideration.

Example 3.10

Calculate the distribution of suspended solids in a reactor for the following treatment systems: (a) conventional activated sludge with a sludge age of 6 days and (b) extended aeration with a sludge age of 22 days. Do not consider the solids in the influent sewage. Assume that the removed BOD_5 load is 100kg/d. Adopt:

- $Y = 0.6\,gVSS/gBOD_5$ removed
- $K_d = 0.09\,d^{-1}$
- VSS/SS ratio in the recently formed solids $= 0.9\,gVSS/SS$
- f_b' (recently formed solids) $= 0.8\,gSS_b/gVSS$

Solution:

a) Conventional activated sludge

- The BOD_5 load removed is equal to $100\,kgBOD_5/d$. Thus:

$$Q.(S_o - S) = 100 \text{ kg/d}$$

- Calculation of f_b (Equation 3.69)

$$f_b = \frac{f_b'}{1 + (1 - f_b').K_d.\theta_c} = \frac{0.8}{1 + (1 - 0.8) \times 0.09 \times 6} = 0.72$$

- *Gross production of volatile suspended solids (Equation 3.72)*

$$P_{xv} \text{ gross} = Y.Q.(S_o - S) = 0.6 \times 100 = 60.0 \text{ kgVSS/d}$$

- *Gross production of total suspended solids (Equation 3.73)*

$$P_x \text{ gross} = P_{xv}/(SSV/SS) = 60.0/0.9 = 66.7 \text{ kgTSS/d}$$

- *Production of inorganic solids (Equation 3.74)*

$$P_{xi} = P_x \text{ gross} - P_{xv} \text{ gross} = 66.7 - 60.0 = 6.7 \text{ kgSS}_i/d$$

- *Gross production of biodegradable suspended solids submitted to a residence time of θ_c (Equation 3.76)*

$$P_{xb} \text{ gross} = (P_{xv} \text{ gross}).f_b = 60.0 \times 0.72 = 43.2 \text{ kgSS}_b/d$$

Example 3.10 (*Continued*)

- *Load of biodegradable suspended solids destroyed (Equation 3.79)*

$$P_{xb} \text{ destroyed} = (P_{xb} \text{ gross}).(K_d.\theta_c)/(1 + f_b.K_d.\theta_c)$$
$$= 43.2 \times (0.09 \times 6)/(1 + 0.72 \times 0.09 \times 6)$$
$$= 43.2 \times 0.39 = 16.8 \text{ kgSS}_b/d$$

- *Net production of biodegradable suspended solids (Equation 3.80)*

$$P_{xb} \text{ net} = P_{xb} \text{ gross} - P_{xb} \text{ destroyed} = 43.2 - 16.8 = 26.4 \text{ kgSS}_b/d$$

- *Production of non-biodegradable volatile suspended solids (Equation 3.77)*

$$P_{xnb} = P_{xv} \text{ gross} - P_{xb} \text{ gross} = 60.0 - 43.2 = 16.8 \text{ kgSS}_{nb}/d$$

- *Net production of volatile suspended solids (Equation 3.81)*

$$P_{xv} \text{ net} = P_{xb} \text{ net} + P_{xnb} = 26.4 + 16.8 = 43.2 \text{ kgVSS/d}$$

- *Net production of total suspended solids (Equation 3.84)*

$$P_x \text{ net} = P_{xv} \text{ net} + P_{xi} = 43.2 + 6.7 = 49.9 \text{ kgTSS/d}$$

- Ratio VSS/TSS (Equation 3.85)

$$\text{VSS/TSS} = (P_{xv} \text{ net})/(P_x \text{ net}) = 43.2/49.9 = 0.87(87\%)$$

- Percentage destruction of the biodegradable suspended solids (Equation 3.86)

$$\% \text{ destruction } X_b = 100.(P_{xb} \text{ destroyed})/(P_{xb} \text{ gross})$$
$$= 100 \times 16.8/43.2 = 39\%$$

- Percentage destruction of the volatile suspended solids (Equation 3.87)

$$\% \text{ destruction } X_v = 100.(P_{xb} \text{ destroyed})/(P_{xv} \text{ gross})$$
$$= 100 \times 16.8/60.0 = 28\%$$

Example 3.10 (*Continued*)

b) Extended aeration

- The BOD_5 load removed is equal to 100 $kgBOD_5/d$ (same as in item a). Thus:

$$Q.(S_o - S) = 100 \text{ kg/d}$$

- Calculation of f_b (Equation 3.69)

$$f_b = \frac{f_b'}{1 + (1 - f_b').K_d.\theta_c} = \frac{0.8}{1 + (1 - 0.8) \times 0.09 \times 22} = 0.57$$

- *Gross production of volatile suspended solids – same as in item a*

$$P_{xv} \text{ gross} = 60.0 \text{ kgVSS/d}$$

- *Gross production of total suspended solids – same as in item a*

$$P_x \text{ gross} = 66.7 \text{ kgTSS/d}$$

- *Production of inorganic suspended solids – same as in item a*

$$P_{xi} = 6.7 \text{ kgSS}_i/d$$

- *Gross production of biodegradable suspended solids submitted to a residence time θ_c (Equation 3.76)*

$$P_{xb} \text{ gross} = (P_{xv} \text{ gross}).f_b = 60.0 \times 0.57 = 34.2 \text{ kgSS}_b/d$$

- *Load of biodegradable suspended solids destroyed (Equation 3.79)*

$$P_{xb} \text{ destroyed} = (P_{xb} \text{ gross}).(K_d.\theta_c)/(1 + f_b.K_d.\theta_c)$$
$$= 43.2 \times (0.09 \times 22)/(1 + 0.57 \times 0.09 \times 22)$$
$$= 34.2 \times 0.93 = 31.8 \text{ kgSS}_b/d$$

- *Net production of biodegradable suspended solids (Equation 3.80)*

$$P_{xb} \text{ net} = P_{xb} \text{ gross} - P_{xb} \text{ destroyed} = 34.2 - 31.8 = 2.4 \text{ kgSS}_b/d$$

- *Production of non-biodegradable volatile suspended solids (Equation 3.77)*

$$P_{xnb} = P_{xv} \text{ gross} - P_{xb} \text{ gross} = 60.0 - 34.2 = 25.8 \text{ kgSS}_{nb}/d$$

- *Net production of volatile suspended solids (Equation 3.81)*

$$P_{xv} \text{net} = P_{xb} \text{net} + P_{xnb} = 2.4 + 25.8 = 28.2 \text{ kgVSS/d}$$

Example 3.10 (Continued)

- *Net production of total suspended solids (Equation 3.84)*

$$P_x \text{ net} = P_{xv} \text{ net} + P_{xi} = 28.2 + 6.7 = 34.9 \text{ kgTSS/d}$$

- *Ratio VSS/TSS (Equation 3.85)*

$$\text{VSS/TSS} = (P_{xv} \text{ net})/(P_x \text{ net}) = 28.2/34.9 = 0.81 \ (81\%)$$

- *Percentage destruction of the biodegradable suspended solids (Equation 3.86)*

$$\% \text{ destruction } X_b = 100.(P_{xb} \text{ destroyed})/(P_{xb} \text{ gross})$$

$$= 100 \times 31.8/34.2 = 93\%$$

- *Percentage destruction of the volatile suspended solids (Equation 3.87)*

$$\% \text{ destruction } X_v = 100.(P_{xb} \text{ destroyed})/(P_{xv} \text{ gross})$$

$$= 100 \times 31.8/60.0 = 53\%$$

c) Summary

Net production (kg/d)	Conventional activated sludge ($\theta_c = 6$ days)	Extended aeration ($\theta_c = 22$ days)
SS biodegradable volatile	26.4	2.4
SS non-biodegradable volatile	16.8	25.8
SS volatile (biodegradable + non biodegradable)	43.2	28.2
SS inorganic	6.7	6.7
SS total	49.9	34.9
Ratio VSS/SS (%)	*87*	*81*
Destruction of biodegradable SS (%)	*39*	*93*
Destruction of volatile SS (%)	*28*	*53*

d) Schematics of the production of the biological solids

See figure below.

e) Comments

- The example demonstrates that the highest production of solids is in systems with lower sludge age (e.g. conventional activated sludge), compared with the production in systems with higher sludge age (e.g. extended aeration).
- If the inorganic and inert solids of the raw sewage had been considered (which is the case, in practice), the total production values would have been different, as well as the VSS/SS ratio.

Example 3.10 (*Continued*)

- The systems with a high sludge age (e.g. extended aeration) lead to a high solids digestion (in this example, 93% removal efficiency of biodegradable suspended solids and 53% removal of the volatile suspended solids produced), compared with systems with a low sludge age (e.g. conventional activated sludge – in this example, 39% efficiency in the removal of biodegradable suspended solids and 28% in the removal of the volatile suspended solids produced).

Production of biological solids (kg/d)
(Example 3.10)

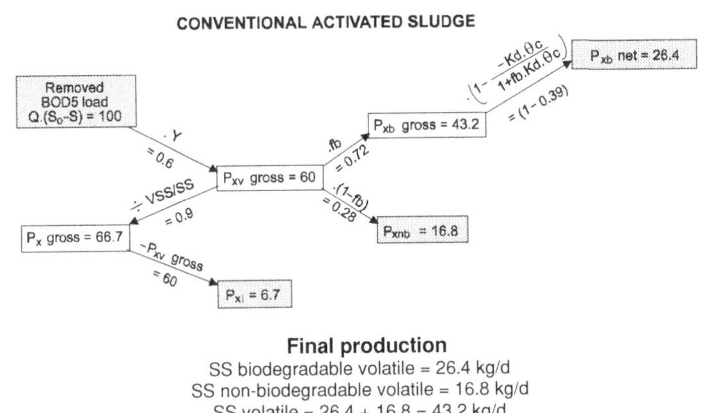

Final production
SS biodegradable volatile = 26.4 kg/d
SS non-biodegradable volatile = 16.8 kg/d
SS volatile = 26.4 + 16.8 = 43.2 kg/d
SS inorganic = 6.7 kg/d
SS total = 43.2 + 6.7 = 49.9 kg/d

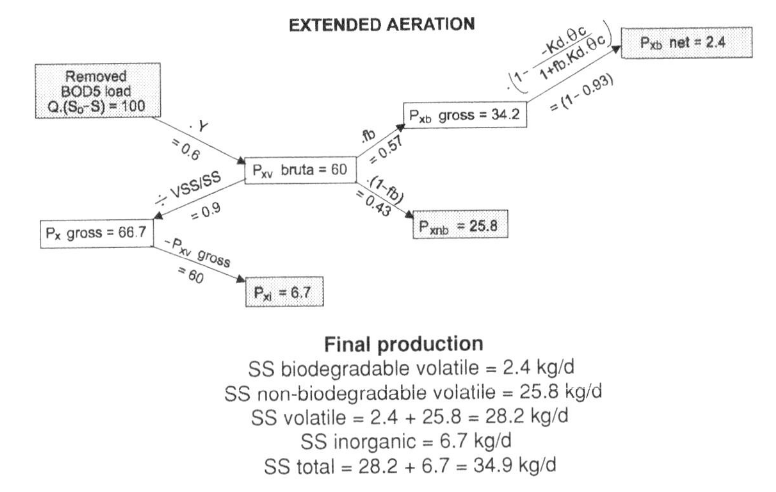

Final production
SS biodegradable volatile = 2.4 kg/d
SS non-biodegradable volatile = 25.8 kg/d
SS volatile = 2.4 + 25.8 = 28.2 kg/d
SS inorganic = 6.7 kg/d
SS total = 28.2 + 6.7 = 34.9 kg/d

4

Sedimentation

4.1 INTRODUCTION

Sedimentation is the physical operation that separates solid particles with a density higher than that of the surrounding liquid. In a tank in which the water flow velocity is very low, the particles tend to go to the bottom under the influence of gravity. As a result, the supernatant liquid becomes clarified, while the particles at the bottom form a sludge layer, and are then subsequently removed with the sludge. Sedimentation is a unit operation of high importance in various wastewater treatment systems.

The main applications of sedimentation in wastewater treatment are:

- **Preliminary treatment**. *Grit removal* (sedimentation of inorganic particles of large dimensions)
  - Grit chamber
- **Primary treatment**. *Primary sedimentation* (sedimentation of suspended solids from the raw sewage)
  - Conventional primary clarifiers, with frequent sludge removal
  - Septic tanks
- **Secondary treatment**. *Secondary sedimentation* (removal of mainly biological solids)
  - Final sedimentation tanks in activated sludge systems
  - Final sedimentation tanks in trickling filter systems
  - Sedimentation compartments in anaerobic sludge blanket reactors
  - Sedimentation ponds, after complete-mix aerated lagoons

- **Sludge treatment**. *Thickening* (settling and thickening of primary sludge and/or excess biological sludge)
  - Gravity thickeners
- **Physical–chemical treatment**. *Settling after chemical precipitation*
  - Enhancement of the performance of primary clarifiers
  - Polishing of effluents from secondary treatment
  - Chemical nutrient removal
  - Physical–chemical treatment (chemical coagulation) of mainly industrial wastewater, but also domestic wastewater

Besides these, sedimentation occurs in various other wastewater treatment units, such as stabilisation ponds, even if they have not been specifically designed for this purpose.

The main objective in most of the applications is to produce a *clarified effluent*, that is, with a low suspended solids concentration. However, at the same time it is also frequently desired to obtain a *thickened sludge* to help its subsequent treatment.

Figures 4.1 and 4.2 present the schematics of two types of settling tanks, one rectangular with horizontal flow, and the other circular with central feeding. Details about the design of these settling tanks are presented in the relevant chapters of this book, related to the various wastewater treatment processes. In the present chapter, only the basic principles of sedimentation are presented.

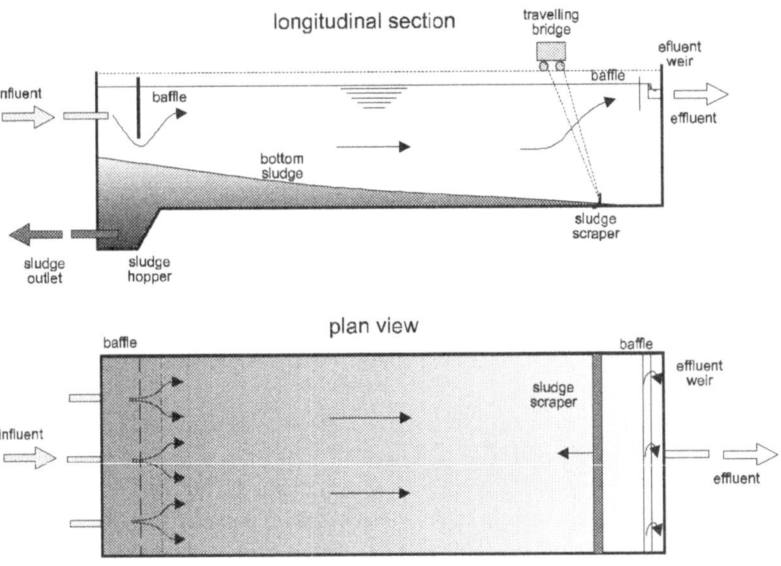

Figure 4.1.   Schematics of a rectangular settling tank with horizontal flow

CIRCULAR SEDIMENTATION TANK

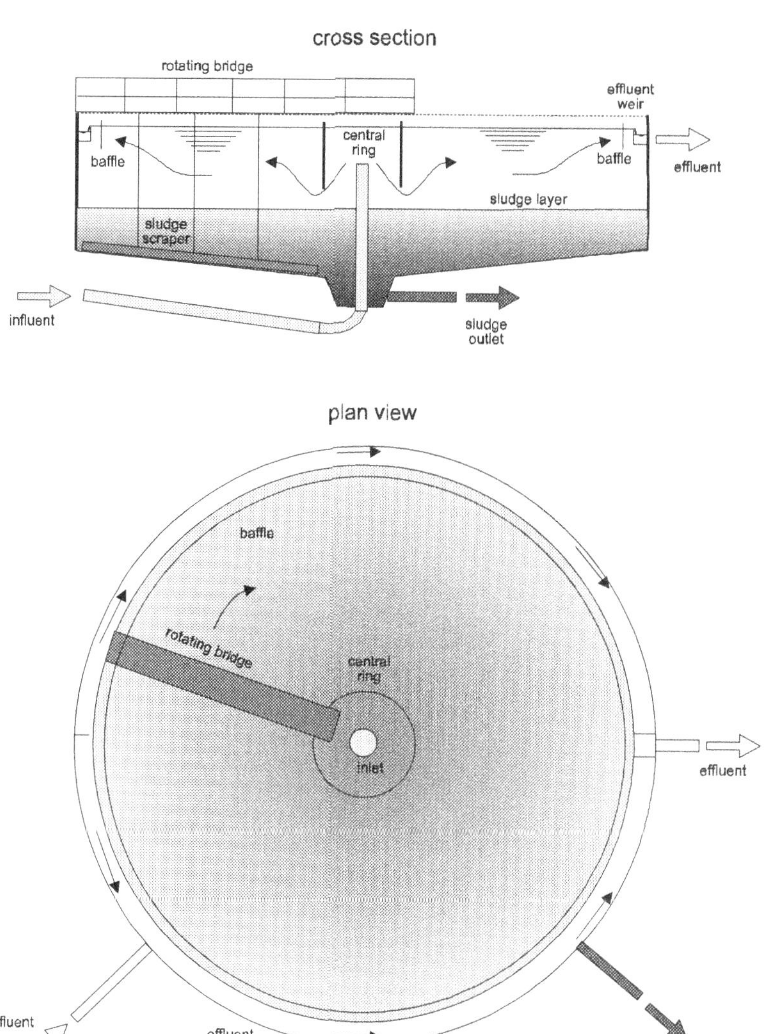

Figure 4.2.   Schematics of a circular settling tank with central feeding

4.2 TYPES OF SETTLING

In wastewater treatment, there are basically the four different types of settling described in Table 4.1. It is probable that during a settling operation more than one type occurs at a given time.

Table 4.1. Settling types in wastewater treatment

Type	Scheme	Description	Example of application/ Occurrence
Discrete	t=0 t=1 t=2	The particles settle, maintaining their identity, that is, they do not coalesce. Hence, their physical properties such as shape, size and density are preserved.	• Grit chambers
Flocculent	t=0 t=1 t=2	The particles coalesce while settling. Their characteristics are changed, with an increase in size (floc formation) and, as a result, in the settling velocity.	• Primary sedimentation tanks • Upper part of secondary sedimentation tanks • Chemical flocs in physical–chemical treatment
Hindered (or zone)	t=0 t=1 t=2	When there is a high concentration of solids, a sludge blanket is formed, which settles as a single mass (the particles tend to stay in a fixed position with relation to the neighbouring particles). A clear separation interface can be observed between the solid phase and the liquid phase. The interface level moves downwards as a result of the settling of the sludge blanket. In this case, it is the settling velocity of the interface that is used in the design of the settling tanks.	• Secondary sedimentation tanks • Sludge gravity thickeners
Compression	t=0 t=1 t=2	If the solids concentration is even higher, the settling could occur only by compression of the particles' structure. The compression occurs due to the weight of the particles, constantly added because of the sedimentation of the particles situated in the supernatant liquid. With the compression, part of the water is removed from the floc matrix, reducing its volume	• Bottom of secondary sedimentation tanks • Sludge gravity thickeners

Source: adapted from Tchobanoglous and Schroeder (1985), Metcalf and Eddy (1991)

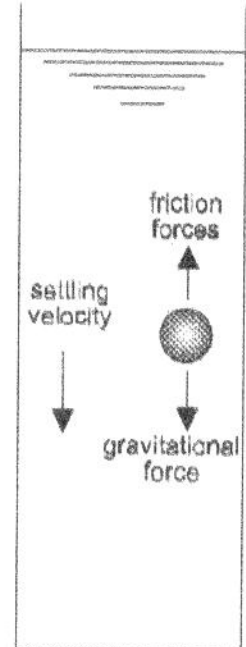

Figure 4.3.   Interacting forces in a particle under discrete settling

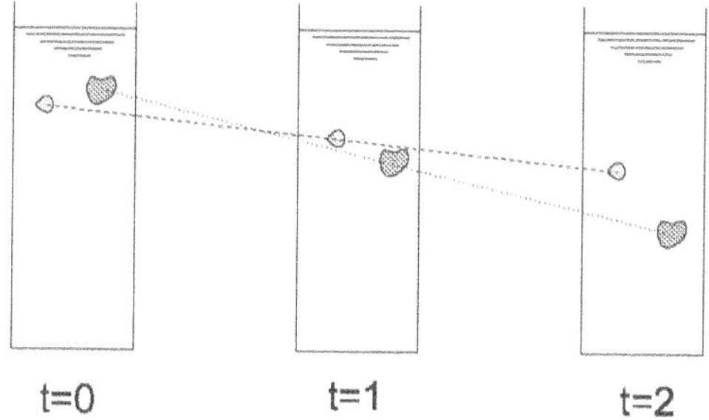

Figure 4.4.   Discrete settling, showing constant settling velocity of the particles

4.3 DISCRETE SETTLING

4.3.1 Settling velocity

The sedimentation of discrete particles can be analysed through the classic laws of Newton and Stokes. According to these laws, the final velocity of a particle under sedimentation in a liquid is constant, that is, the frictional force is equal to the gravitational force. This terminal velocity is reached in the liquid medium in fractions of a second. Figure 4.3 shows the intervening forces on a settling particle, while Figure 4.4 emphasises the fact that the settling velocity of discrete particles is constant.

According to Stokes law, the discrete settling velocity of a particle (v_s) in laminar flow is given by:

$$v_s = \frac{1}{18} \cdot \frac{g}{\upsilon} \cdot \frac{\rho_s - \rho_l}{\rho_l} \cdot d^2 \qquad (4.1)$$

where:

v_s = settling velocity of the particle (m/s)
g = acceleration due to gravity (m/s^2)
υ = kinematic viscosity of the liquid (m^2/s)

ρ_s = particle density (kg/m^3)
ρ_l = liquid density (kg/m^3)
d = particle diameter (m)

The kinematic viscosity v and the density of water ρ_l are functions of the temperature T. However, the variation in the density of the water within the usual temperature ranges in wastewater treatment can be neglected (999.8 kg/m^3 and 992.2 kg/m^3 for temperatures of 0 °C and 40 °C, respectively) and a value of 1000 kg/m^3 can be adopted. The influence on the water viscosity is more representative, as seen in Table 4.2 (Tchobanoglous and Schroeder, 1985; Huisman, 1978).

Table 4.2. Kinematic viscosity of the water as a function of temperature

T (°C)	0	5	10	15	20
v(m^2/s)	1.79×10^{-6}	1.52×10^{-6}	1.31×10^{-6}	1.15×10^{-6}	1.01×10^{-6}
T (°C)	25	30	35	40	
v(m^2/s)	0.90×10^{-6}	0.80×10^{-6}	0.73×10^{-6}	0.66×10^{-6}	

In the range of T = 10 to 30 °C, von Sperling (1999) proposes the following equation for the viscosity as a function of the temperature ($R^2 = 0.986$):

$$v = 3.76 \times 10^{-6} \times T^{-0.450} \tag{4.2}$$

When interpreting Equation 4.1, the following considerations are important:

- v_s *is proportional to* $(\rho_s - \rho_l)/\rho_l$
- v_s *is proportional to* d^2

The fact that v_s is proportional to the square of the particle diameter emphasises the importance of the increase in the size of the particles, aiming at a faster particle removal, and, consequently, smaller sedimentation tanks. As an example, when the particle diameter doubles, the settling velocity increases four times.

Example 4.1

Calculate the settling velocity of a sand grain using the following data:

- Grain diameter: d = 0.7 mm
- Sand density: ρ_s = 2650 kg/m^3
- Liquid density: ρ_l = 1000 kg/m^3
- Liquid temperature: T = 25 °C

Solution:

From Table 4.2, for the temperature of 25 °C, the kinematic viscosity of the water v is 0.90×10^{-6} m^2/s. The diameter of the particle is 0.7×10^{-3} m. From Equation 4.1, assuming laminar flow:

$$v_s = \frac{1}{18} \cdot \frac{g}{v} \cdot \frac{\rho_s - \rho_l}{\rho_l} \cdot d^2 = \frac{1}{18} \cdot \frac{9.81}{0.90 \times 10^{-6}} \cdot \frac{2650 - 1000}{1000} \cdot (0.7 \times 10^{-3})^2$$

$$= 0.49 \text{ m/s}$$

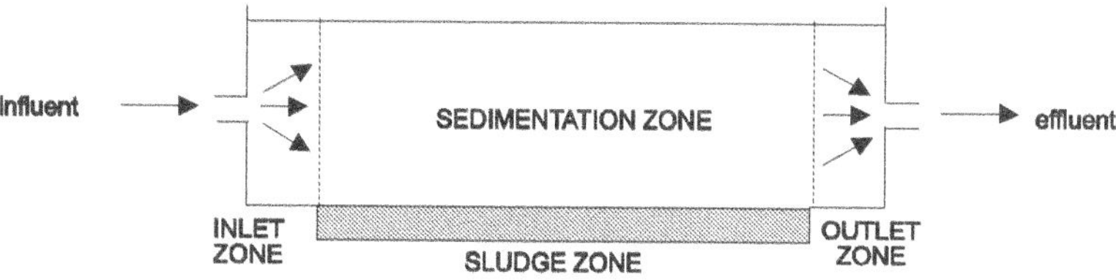

Figure 4.5. Schematic representation of the zones in a horizontal sedimentation tank (longitudinal section)

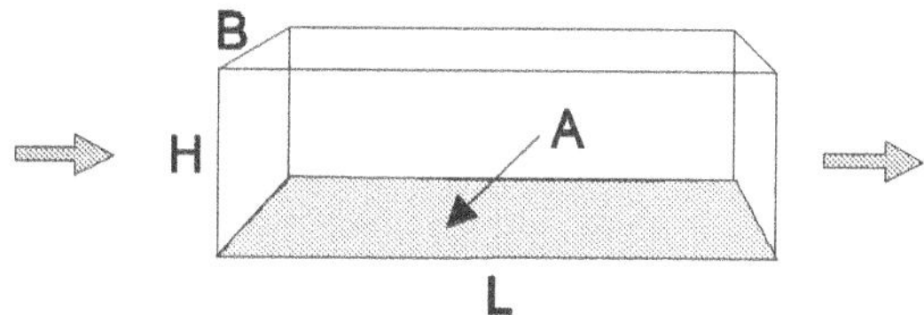

Figure 4.6. Dimensions in the sedimentation zone

4.3.2 The concept of an ideal sedimentation tank with horizontal flow

The discrete settling of a particle can be analysed in a settling column without flow as well as in a rectangular horizontal-flow tank with constant horizontal velocity (v_h). Figure 4.5 shows the representative zones of this ideal tank. The theoretical considerations apply to the zone where settling effectively occurs (sedimentation zone).

For the theoretical analysis of sedimentation, it is necessary to assume that:

- the particles are uniformly distributed in the inlet zone
- the particles that touch the sludge zone are considered removed
- the particles that reach the outlet zone are not removed by sedimentation

The main dimensions of the *sedimentation zone* are presented in Figure 4.6.

In an ideal sedimentation tank with constant horizontal velocity, the discrete settling of a particle occurs as in a sedimentation column (see Figure 4.7).

The *time* taken for a particle to reach the bottom is given by:

- *sedimentation column*: time = distance/velocity

$$t = \frac{H}{v_s} \qquad (4.3)$$

- *horizontal flow tank*: time = volume/flow

$$t = \frac{V}{Q} = \frac{H.A}{Q} \qquad (4.4)$$

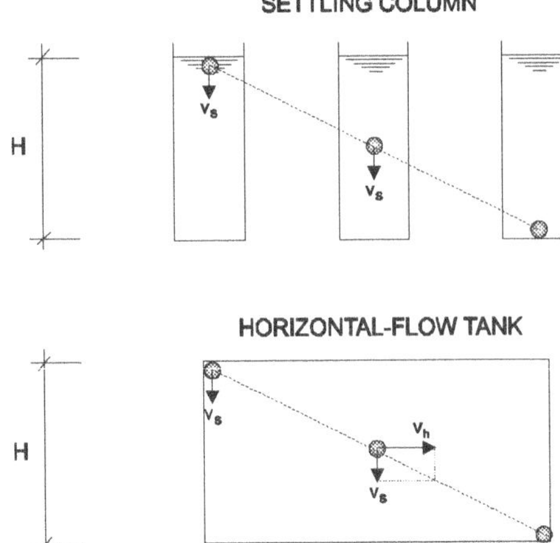

Figure 4.7.   Discrete settling of a particle in a sedimentation column and in a horizontal tank

Combining Equations 4.3 and 4.4:

$$v_s = \frac{Q}{A}$$

(4.5)

This equation is very important in the design of sedimentation tanks. If it is desired to remove particles with settling velocities equal to or greater than v_s, and knowing the wastewater flow to be treated Q, the required surface area can be obtained from:

$$A = \frac{Q}{v_s}$$

(4.6)

The settling velocity to be adopted for design (v_s, or v_o) is also called *overflow rate* or *hydraulic surface loading rate*, and is expressed in units of velocity (m/h), or flow per unit area ($m^3/m^2.h$).

In the interpretation of Equation 4.5, it should be noted that:

- v_s can be obtained through experiments with the liquid to be treated or from literature values (in a design, v_s is a design parameter)
- the removal of discrete particles depends only on the surface area (A) and not on the height (H) and time (t).

The last point can be understood as follows. If A and Q are maintained constant, and if H doubles, the volume V doubles, and so does the time t (see Equation 4.4). The horizontal velocity v_h ($v_h = Q/(B.H)$) is reduced to half. Since v_s is constant (function only of the particle characteristics), the new trajectory of the particle

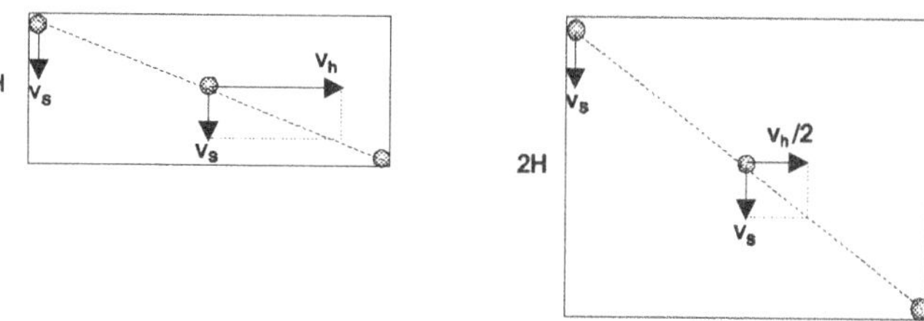

Figure 4.8. Visualisation of the non-influence of H on the removal of discrete particles.

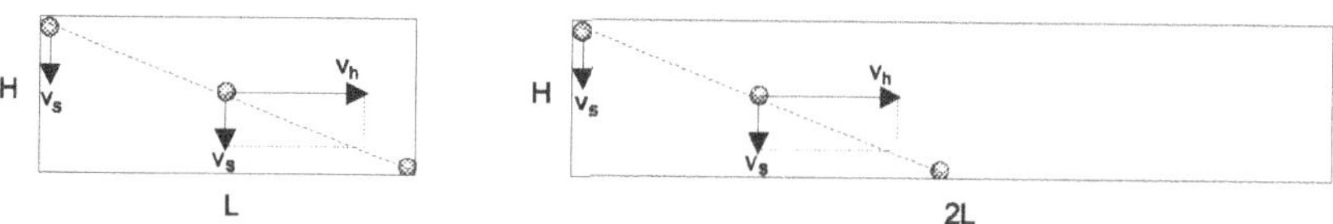

Figure 4.9. Visualisation of the influence of A on the removal of discrete particles

leads to its removal in the final extremity of the tank, identically to the tank with a lower height (see Figure 4.8).

However, if the surface area A doubles, for example through the duplication of the length L, v_h, and v_s remain constant. The trajectory of the particle is not altered, but the particle is removed in half of the tank length (see Figure 4.9). Hence, this new tank is able to receive particles with settling velocities lower than v_s. In summary, *for the **ideal discrete settling**, the surface area A is of fundamental importance, while H and t do not play any role.*

The particles to be removed in a sedimentation tank depend on the:

- settling velocity of the particle (compared with the design settling velocity v_s)
- height at which the particle enters the sedimentation zone

In a tank removing particles originating from different vertical positions and with different settling velocities there are the possibilities shown in Table 3.2.

4.3.3 Discrete settling tests

A large diversity of particle sizes occurs in a typical suspension of particulate matter. To determine the removal efficiency in a certain time, it is necessary to consider the whole range of the settling velocities found in the system. This is usually done through tests in a settling column, in which samples are extracted from various depths and times (Metcalf & Eddy, 1991).

The settling column (see Figure 4.10) needs sampling points at various levels. Typical analyses made with the samples are suspended solids (SS), which allow

Table 4.3. Discrete particles to be removed in a horizontal flow tank

Case	Particles removed or not removed
CASE 1 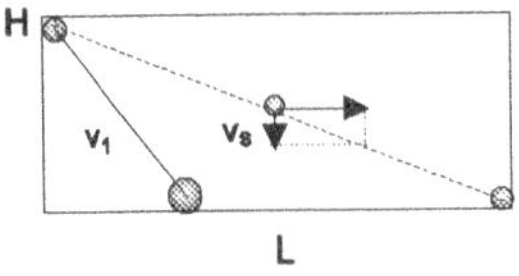	*Particles removed:* • particles with a settling velocity equal to v_s that enter the tank at a height H • particles with a settling velocity $v_1 > v_s$ that enter the tank at a height H
CASE 2 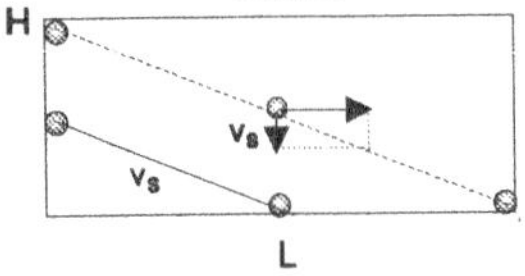	*Particles removed:* • particles with a settling velocity equal to v_s that enter the tank at a height lower than H
CASE 3 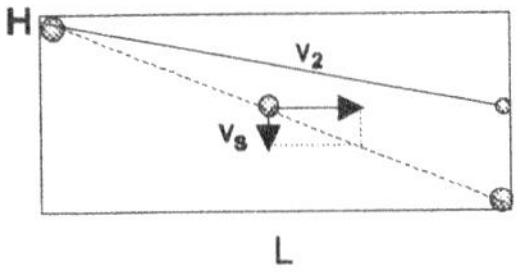	*Particles <u>not</u> removed:* • particles with a settling velocity $v_2 < v_s$ that enter the tank at a height H
CASE 4 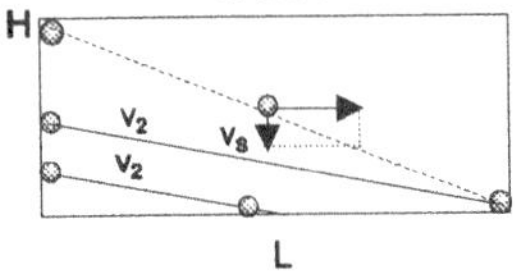	*Particles that <u>can</u> be removed:* • particles with a settling velocity $v_2 < v_s$ that enter the tank at a height lower than H
CASE 5 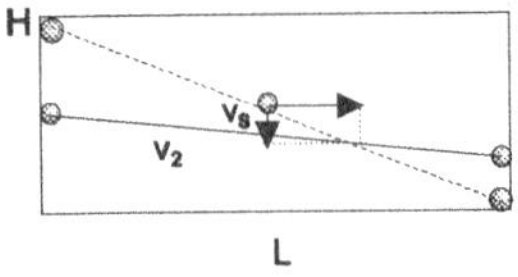	*Particles that <u>may be not</u> removed:* • particles with a settling velocity $v_2 < v_s$ that enter the tank at a height lower than H

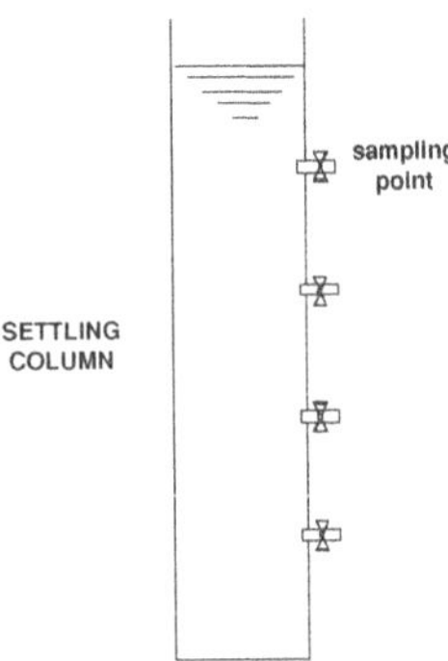

Figure 4.10. Settling column

the evaluation of the removal efficiency of the particulate matter. At the start of the test, the column should be full of a homogeneous mixture of the suspension. The samples are taken at different times, at the various sampling points along the column height. If the initial concentration in the column at the initial time $t_o = 0$ is C_o, and after a time t_i, the concentration reduces to C_i at the depth z_i, then $C_o - C_i$ of the original suspension has settling velocities greater then $z_i / (t_i - t_o)$. Repeating this concept for different depths and times, the cumulative curve of the proportion of particles with settling velocities lower than the X-axis value can be constructed (Wilson, 1981).

Example 4.2 (adapted from Wilson, 1981) presents a methodology for the determination of the removal efficiency of discrete particles based on a settling test.

Example 4.2

The results of a settling test done in a suspension with particles that present discrete settling led to the values presented below. Plot the cumulative profile of the settling velocity and calculate the fraction of particles removed for an overflow rate of $v_o = 1.0$ m/h.

Sample	Sampling depth (m)	Sampling time (h)	SS in the sample (mg/L)
1	0.0	0.0	
2	0.0	0.0	222 (average)
3	0.0	0.0	
4	1.0	1.0	140
5	1.0	3.0	108
6	1.0	6.0	80
7	2.0	1.0	142
8	2.0	3.0	110
9	2.0	6.0	106
10	3.0	1.0	142
11	3.0	3.0	130
12	3.0	6.0	124
13	4.0	1.0	147
14	4.0	3.0	126
15	4.0	6.0	114

Solution:

a) Plot the curve of the fraction of particles $\times$ settling velocity

Consider *sample 9* (depth of 2.0 m and sample time of 6.0 h). The settling velocities of the particles found in the sample are less than 2.0 m/6.0 h = 0.33 m/h. Hence, $(106/222) = 0.48 = 48\%$ of the particles have a settling velocity lower than 0.33 m/h. The removal efficiency in this sample is $1 - 0.48 = 0.52 = 52\%$. With this reasoning, the table below can be constructed.

Example 4.2 (Continued)

Sample	Velocity (m/h)	Fraction of the SS remaining
4	1.00	0.63
5	0.33	0.49
6	0.17	0.36
7	2.00	0.64
8	0.67	0.50
9	0.33	0.48
10	3.00	0.64
11	1.00	0.59
12	0.50	0.56
13	4.00	0.66
14	1.33	0.57
15	0.67	0.51

Based on the values from the table above, a graph of the settling velocities × remaining SS fraction can be constructed. Note that the sampling depth has no influence on the discrete settling test.

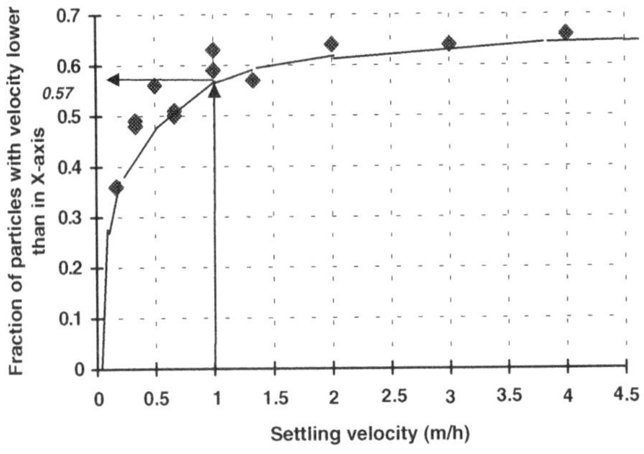

b) Determination of the fraction of particles removed

From the figure above, 0.57 (57%) of the particles have a settling velocity lower than 1.0 m/h. Thus, the fraction removed of these particles, if they started settling from the top of the column, would be $1 - 0.57 = 0.43$ (43%).

There is still another fraction removed, which corresponds to the particles with a settling velocity lower than v_o, which did not start settling from the top of the column (or the top of the horizontal tank). The removal fraction of these particles is given by the area occupied between the Y-axis and the curve until $x = 3.0$ m/h. This can be obtained from the table below, which presents the

---

Example 4.2 (*Continued*)

calculation of the area based on a division into strips, with the width (dx_i) and the average velocity in the strip (vx_i):

Strip of dx_i *(Y axis)*	Width of the strip (dx_i) *(Y axis)*	Average velocity in the strip (vx_i) (m/h) *(X axis)*	$dx_i.vx_i$
0.50–0.57	0.07	0.80 (at y = 0.54)	0.056
0.40–0.50	0.10	0.36 (at y = 0.45)	0.036
0.30–0.40	0.10	0.14 (at y = 0.35)	0.014
0.20–0.30	0.10	0.05 (at y = 0.25)	0.005
0.10–0.20	0.10	$\simeq$ 0 (at y = 0.15)	–
0.00–0.10	0.10	$\simeq$ 0 (at y = 0.05)	–
Total	–	–	0.111

The fraction removed in this second way is:

$$\frac{\sum dx_i.vx_i}{v_0} = \frac{0.111}{1.0} = 0.11$$

The total fraction removed is $0.43 + 0.11 = 0.54$ (54%).

Therefore, for the overflow rate of 1.0 m^3/m^2.h, 54% of the particles in the sampled suspension are removed.

If desired, a curve showing the removal efficiencies as a function of the overflow rate (v_0) can be constructed, based on a repetition of the calculations for different values of v_0.

4.4 FLOCCULENT SETTLING

In flocculent settling, the particles agglomerate themselves and form flocs that tend to grow in size while settling. With the increase in the size of the particles (flocs), there is an increase in the settling velocity. Therefore, in flocculent settling the velocity is not constant as in discrete settling, but tends to increase. Figures 4.11 and 4.12 show the floc formation process and the increase in the settling velocity in a horizontal flow tank (Figure 4.11) and in a settling column (Figure 4.12).

Since the flocculation occurs while the particles go to the bottom, the greater the chance of contact they have, the greater will be the floc formation. As a result, *the removal efficiency in flocculent settling is increased with an increase in the depth H and in the time t* (differently from discrete settling).

Similarly to discrete settling, flocculent settling in an ideal horizontal flow tank can be compared with settling in a column without flow.

In the case of flocculent settling, the settling velocity of the individual particles is not analysed, as in the case of discrete settling. Settling column tests are also

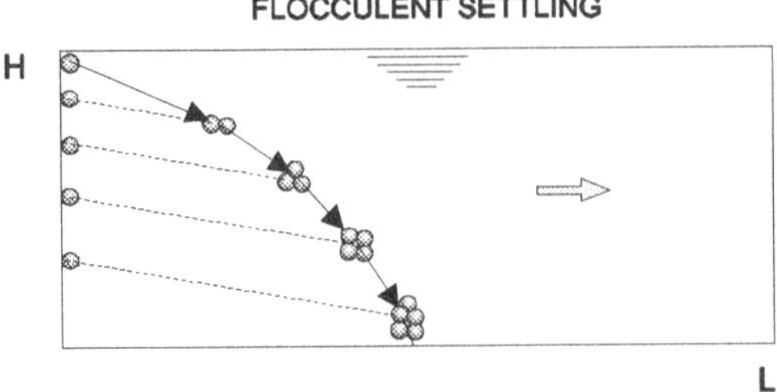

Figure 4.11.   Flocculent settling in a horizontal flow tank

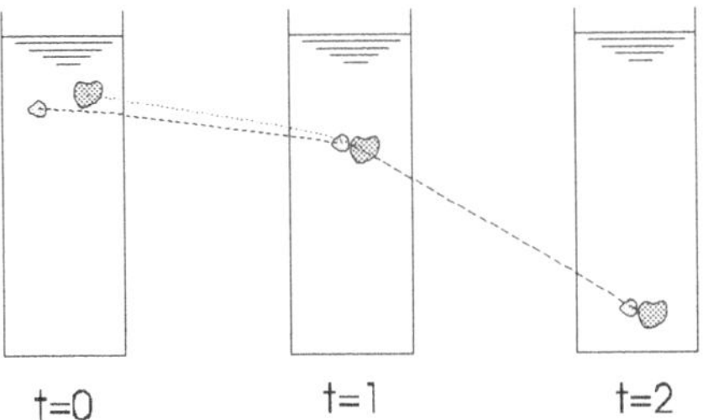

Figure 4.12.   Flocculent settling in a settling column

useful here to permit the selection of the ideal overflow rate v_0. In the flocculent settling test, the results are presented in the form of curves or a grid, showing the particle removal percentages at certain depths and times (see Example 4.3, adapted from Wilson, 1981).

Example 4.3

Assuming the values presented in the table below, calculate the expected percentage removal in the following cases:

- 2.0 m deep tank, with a detention time of 1.50 h
- 2.0 m deep tank, with a detention time of 3.00 h
- 1.0 m deep tank, with a detention time of 3.00 h
- 2.5 m deep tank, with a detention time of 3.75 h

Percentage removal (%)

Sampling depth (m)	Sampling time (h)							
	0.5	1.0	1.5	2.0	2.5	3.0	3.5	4.0
1.0	26	44	49	55	63	66	71	77
2.0	20	34	44	51	56	60	62	64
3.0	19	27	37	45	51	57	60	68

Example 4.3 (Continued)

Solution:

The percentage removal efficiencies are calculated based on the initial concentration C_o, and on the concentration at the sampling time. For example, if the initial concentration were $C_o = 200$ mg/l, and the concentration at the depth and time of sampling is 132 mg/l, the removal efficiency would be $(200-132)/200 = 0.34 = 34\%$. This could have been the case, for example for a sample at a 2.0m depth and 1.0h of sampling.

a) Tank with 2.0 m depth and detention time of 1.50 h

For the depth of 2.0 m and sampling time of 1.5 h, the percentage removal can be obtained directly from the table. The value found is 44%.

b) Tank with 2.0 m depth and detention time of 3.0 h

For the depth of 2.0 m and sampling time of 3.0 h, the percentage removal shown in the table is 60%. The depth is the same as in item a, but the detention time (or the volume) is the double. Since both depths are the same, but the volumes are different, then the surface area of item b is the double of that in item a. The comparison with the calculation of item a, in terms of design, is that the maintenance of the same depth of 2.0 m but with a duplication of the volume, surface area or detention time (3.0 h compared with 1.5 h) leads to an increase of the removal efficiency from 44% to 60%.

c) Tank with 1.0 m depth and detention time of 3.0 h

For the depth of 1.0 m and sampling time of 3.0 h, the percentage removal expressed in the table is 66%. The detention time (or the volume) is the same as item b, but the depth is half of that. Since both volumes are equal, but the depth is different, then the surface area of item c is double that of item b. A comparison with the calculation of item b, in terms of design, is that the maintenance of the same volume but with a reduction in the depth and duplication of the surface area leads to an increase of the removal efficiency from 60% to only 66%.

d) Tank with 2.5 m depth and detention time of 3.75 h

For the depth of 2.5 m and sampling time of 3.75 h, an interpolation in the grid is necessary. Adopting a linear interpolation:

$$\% \text{ removal} = \frac{1}{2} \cdot \left(\frac{64+62}{2} + \frac{60+68}{2} \right) = 63.5\%$$

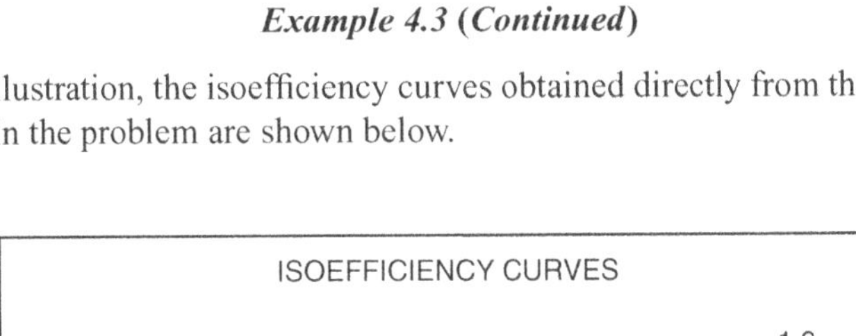

Example 4.3 (Continued)

As an illustration, the isoefficiency curves obtained directly from the data presented in the problem are shown below.

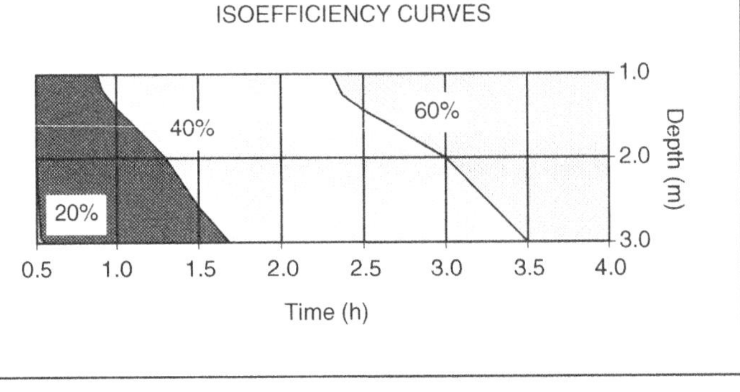

4.5 ZONE SETTLING

4.5.1 Settling in a column

When there is a high solids concentration, a blanket tends to form. This blanket settles as a single mass of particles (the particles tend to remain in a fixed position with relation to the neighbouring particles). A clear separation interface can be observed between the solid phase and the liquid phase, and the level of the interface moves downwards as a result of the sedimentation of the sludge blanket (see Figure 4.13). For the blanket to move downwards, the liquid situated underneath tends to move upwards. In the zone settling, it is the settling velocity of the interface that is used in the design of sedimentation tanks. *Zone settling* is also called *hindered settling*.

In a *settling column* completely homogenised with a high concentration of suspended solids, under quiescent conditions and after a short time, a clear interface

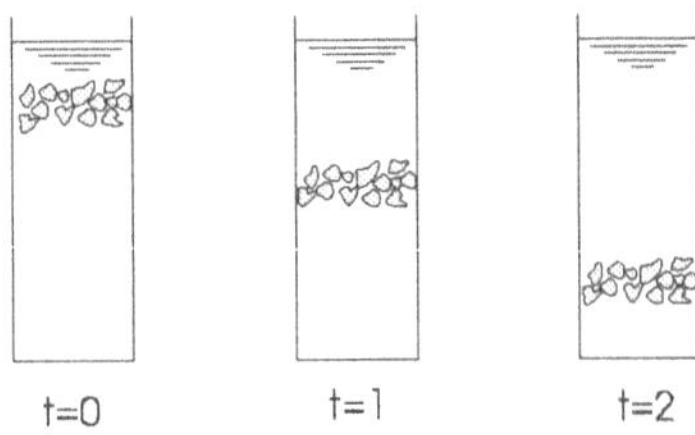

Figure 4.13.   Zone settling of a solids mass in a settling column

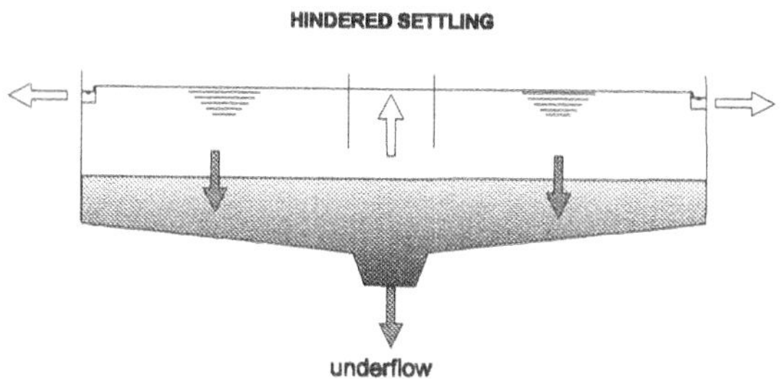

Figure 4.14.   Sludge removal from the bottom in secondary sedimentation tanks

is formed. While the interface moves downwards, the supernatant liquid becomes clarified, and a layer with a higher concentration is formed at the bottom. The level of this highly concentrated layer moves upwards due to the continuous increase of the accumulated material at the bottom, which cannot leave the column from its bottom.

In a *sedimentation tank* with continuous withdrawal of the settled sludge from the bottom, the more concentrated layer does not propagate upwards. The reason is that the underflow velocity of the sludge (downward, from the bottom) counterbalances the expansion velocity (upwards). This situation occurs in tanks with continuous sludge removal from the bottom, such as secondary sedimentation tanks in the activated sludge process (see Figure 4.14).

Figure 4.15 presents schematically the behaviour of the layers created in these two distinct conditions (*without* and *with* sludge removal from the bottom).

4.5.2 The limiting solids flux theory

The solids flux theory describes the zone settling phenomenon that takes place in secondary sedimentation tanks and gravity thickeners. The solids flux theory is a result of the sequential development from many authors, but achieved a greater applicability in the context of wastewater treatment based on the works of Dick (1972). Its utilisation can be for design as well as for operational control. Within a global view of the treatment system, the theory can be used together with a mathematical model of the reactor in order to allow an optimal design of the system (Keinath et al, 1977; Catunda and van Haandel, 1987) or its optimal control (von Sperling, 1990).

The following items focus on the behaviour of a secondary sedimentation tank in an activated sludge plant, due to their greater importance when compared with gravity thickeners. However, the general principles are the same in both cases.

In this context, *flux* can be understood as the solids load per unit area (for example expressed as $kgSS/m^2.h$). In a continuous flow sedimentation tank

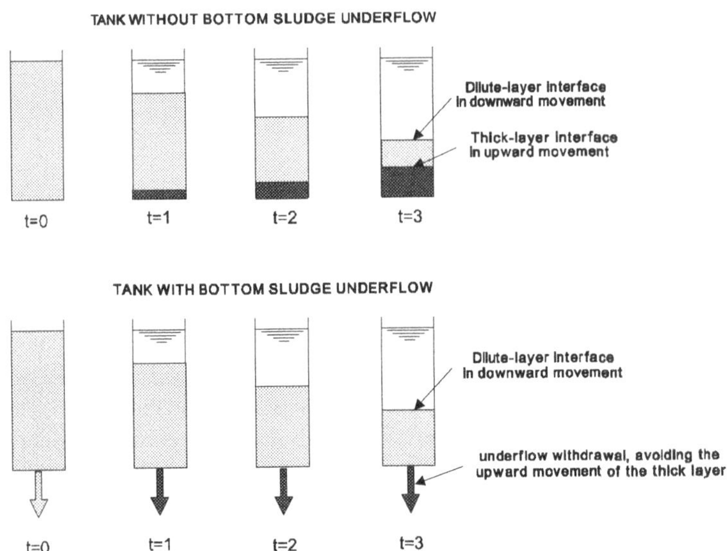

Figure 4.15.   Schematic representation of the behaviour of the diluted and concentrated sludge layers in the zone settling. (a) Column without sludge removal from the bottom. (b) Settling tank with sludge removal from the bottom.

the solids tend to go to the bottom due to the simultaneous action of two fluxes:

- *gravity flux* (G_g), caused by the gravitational sedimentation of the sludge;
- *underflow flux* (G_u), caused by the downward movement originating from the removal of the return sludge from the bottom of the sedimentation tank.

The total flux (G_t) moving to the bottom of the sedimentation tank corresponds to the sum of these two components. The mathematical representation of these fluxes can be expressed as (Dick, 1972):

Total flux:

$$G_t = G_g + G_u \qquad (4.7)$$

Gravity flux:

$$G_g = C.v \qquad (4.8)$$

Underflow flux:

$$G_u = C.\frac{Q_u}{A} \qquad (4.9)$$

where:
C = concentration of suspended solids in the sludge (kg/m^3)
v = settling velocity of the interface at the concentration C (m/h)
Q_u = underflow from the bottom (m^3/h)
A = surface area of the sedimentation tanks (m^2).

The settling velocity v is, on the other hand, a function of the concentration C itself, decreasing with the increase of C. There are various empirical relations to express v in function of C, but the most frequently used is:

$$v = v_0.e^{-K.C} \qquad (4.10)$$

where:
v_0 = coefficient, expressing the interface settling velocity at a concentration
　　C = 0 (m/h)
K = settling coefficient (m^3/kg)

The flux of solids conveyed to the bottom of the sedimentation tank depends on the concentration C, according to the following conditions:

- **Low concentration of C.** With low values of **C**, the settling velocity of the interface **v** is high (Equation 4.10), but the product **C.v** is low, which results in a low value of the gravity solids flux (Equation 4.8).
- **Intermediate concentration of C.** While **C** increases, even with the decrease in **v**, the product **C.v** increases, that is, the gravity flux increases.
- **High concentration of C.** However, after a certain value of C, the reduction in the settling velocity **v** is such that the product **C.v** starts to decrease.

Figure 4.16a presents the curve of the gravity solids flux (G_g = C.v). The intercept of the straight line with slope Q_u/A, tangent to the descending reach of the flux curve, with the Y-axis characterises the **limiting flux** (G_L). This can be understood as the maximum flux that can be transported to the bottom of the sedimentation tank with the existing settleability, sludge concentration and underflow. The same interpretation can be obtained from Figure 4.16b, in which the total flux ($G_t = G_g + G_u$) is presented. The result is the same, but in this case, the limiting flux is obtained at the minimum of the total flux curve. This point of minimum indicates that, while the solids concentration increases in the settling tanks from the inlet to the bottom, there will be a concentration (limiting concentration C_L) that will bring about the lowest flux (limiting flux G_L). At this point, the settling tank is limited and cannot transmit to the bottom a quantity of solids higher than the limiting value. The construction of the solids flux curves is presented in Sections 4.5.3 and 4.5.5.

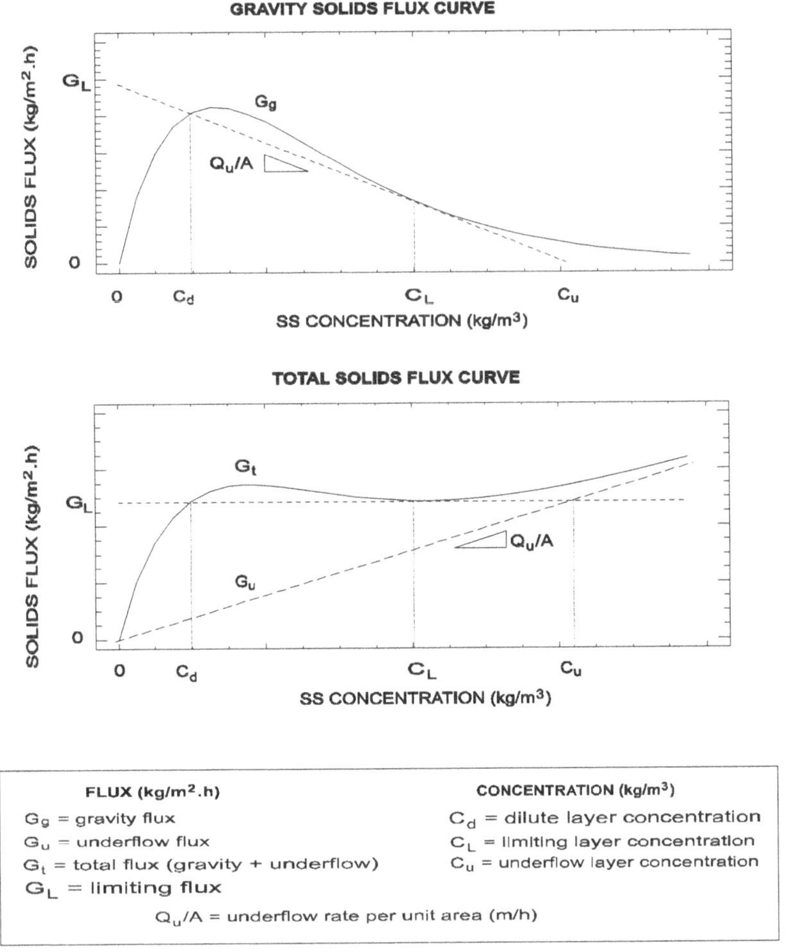

Figure 4.16.　Solids flux curves. (a) Gravity solids flux. (b) Total solids flux.

The success of the design and operation of secondary sedimentation tanks depends on the relation between the *applied flux* and the *limiting flux*. The applied flux corresponds to the load of the influent suspended solids to the settling tank per unit surface area, given by (see Figure 4.17):

$$G_a = \frac{Q_i + Q_r}{A} \cdot C_o \qquad (4.11)$$

where:
G_a = applied solids flux (kg/m².h)

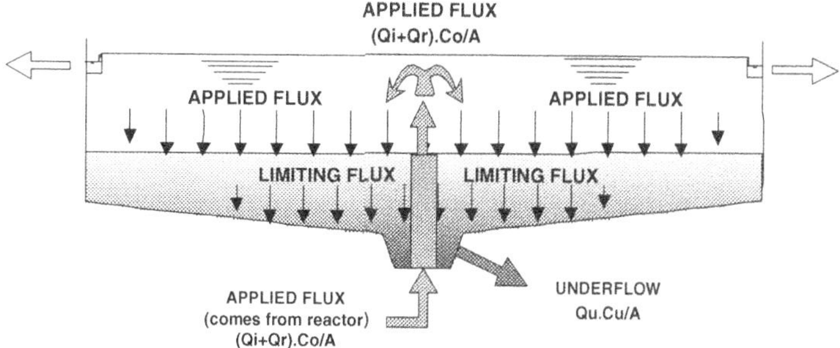

Figure 4.17. Applied solids flux on a secondary sedimentation tank

Q_i = influent flow to the sewage treatment plant (m^3/h)

Q_r = return sludge flow (approximately equal to the sludge underflow) (m^3/h)

C_o = concentration of influent suspended solids to the secondary sedimentation tank. Equal to the suspended solids concentration in the reactor, or mixed liquor suspended solids concentration (MLSS) (kg/m^3).

In practical terms, it can be considered that Q_r is equal to Q_u, since the excess sludge flow ($Q_u - Q_r$) is negligible in the mass balance on the secondary sedimentation tank.

In broad terms, the applied flux must be equal to or less than the limiting flux ($G_a \leq G_L$), so that the settling tank does not accumulate solids, which could eventually reach a quantity that would lead to their loss in the supernatant of the settling tank, thus deteriorating the final effluent quality. On the curve of the gravity flux (Figure 4.16a), the straight line of the applied flux can be drawn. This line starts at the Y-axis (at the value G_a) and goes downwards, in parallel with the line of the limiting flux (slope equal to Q_u/A).

Figure 4.18 presents an interpretation, by the author, of the theory presented by Dick (1972), Handley (1974), White (1976) and Keinath (1981), in terms of the relation between the gravity flux curve and the profile of the suspended solids concentration in the secondary sedimentation tank. In this figure, there are four columns, representing four distinct conditions: (a) *sedimentation tank with underload*, (b) *sedimentation tank with critical load*, (c) *sedimentation tank with thickening overload* and (d) *sedimentation tank with thickening and clarification overload*. The first line presents the curve of the **gravity flux** (equal in the four columns), the **limiting flux** (also equal in the four columns) and the **applied flux** (different in the four columns). The second line presents the vertical profile of the suspended solids concentration resulting from the interrelation between the applied and the limiting fluxes. The third line shows this profile as a cross-section in the settling tank.

SOLIDS FLUX CURVES AND SUSPENDED SOLIDS CONCENTRATION PROFILES

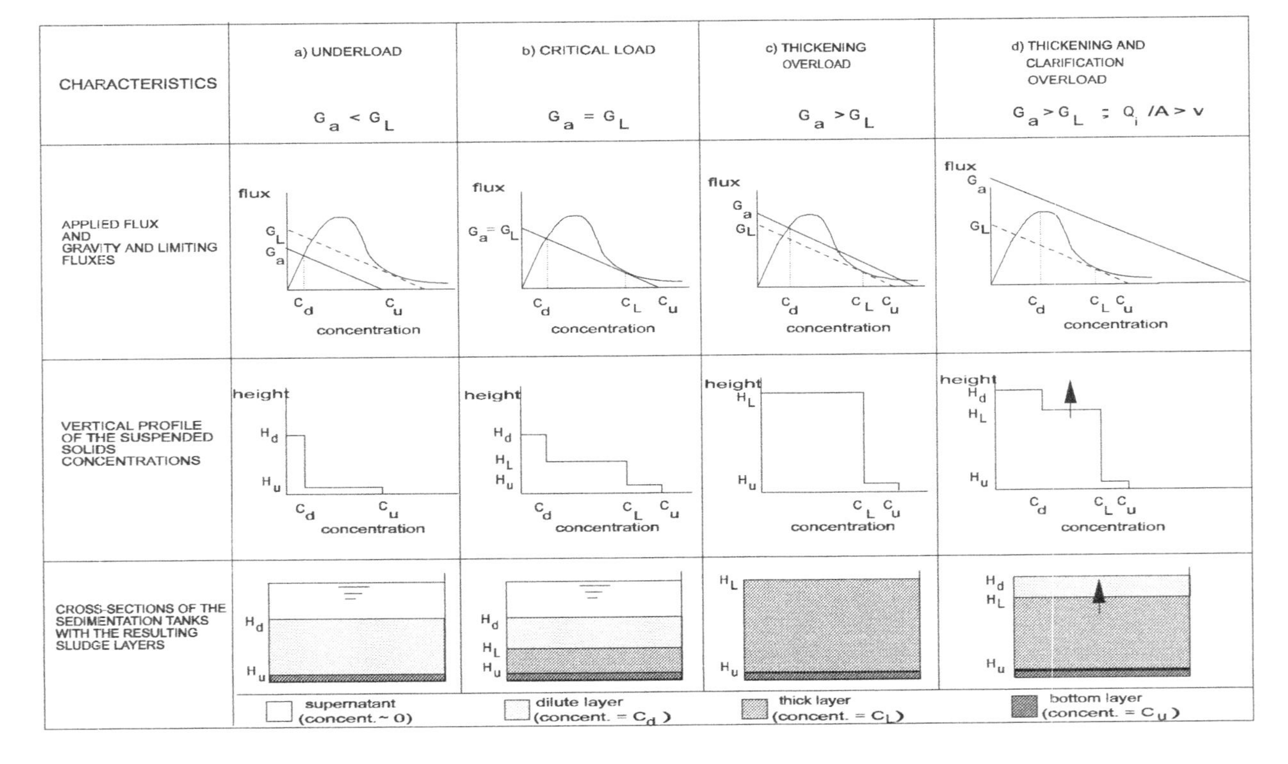

Figure 4.18.   Solids flux curves and suspended solids concentration profiles in secondary sedimentation tanks, for various relations between the applied and the limiting fluxes.

The following points are associated with the interpretation of Figure 4.18:

- **Sedimentation tank with underload**. The settling tank will be underloaded when the applied flux is less than the limiting flux. In this condition, only a diluted layer with a low suspended solids concentration (C_d) will be formed. At the bottom of the settling tank a layer with concentration C_u (concentration of the sludge removed) will also develop, due to the support by the tank bottom.
- **Sedimentation tank with critical load**. The settling tank will be critically loaded when the applied flux is equal to the limiting flux. In this case, a thicker sludge layer (concentration C_L) will be formed.
- **Sedimentation tank with thickening overload**. The settling tank will be overloaded in terms of thickening of the sludge when the applied flux is greater than the limiting flux. In this condition, the concentration of the thick sludge layer will not go beyond C_L and, consequently the thick sludge layer will increase in volume, propagating upwards. Depending on the level reached by the sludge blanket, solids may be discharged with the final effluent.
- **Sedimentation tank with thickening and clarification overload**. The settling tank will be overloaded in terms of thickening and clarification when, besides having an applied flux greater than the limiting flux, the overflow rate (Q_i/A) is greater than the sludge settling velocity v. In this case, the diluted layer as well as the thick layer will propagate upwards, with a possible even faster deterioration of the effluent quality.

4.5.3 Determination of the interface settling velocity

The settling velocity of the interface, also called *zone settling velocity* (ZSV) can be determined experimentally through tests in a settling column. For this, the following simplified methodology is suggested:

- homogenise the liquid through mixing in the whole tank (column) volume
- interrupt mixing to allow sedimentation
- measure the interface level at various time intervals
- stop the measurement when the interface is not significantly settling any more
- plot a graph: interface height (Y axis) × time (X axis)
- determine the settling velocity of the interface by the slope of the straight-line reach in the graph (ignore the initial and final points that are not on the straight-line reach)

The test is commonly done in cylinders up to 0.5 m in height and 10 cm in diameter. However, whenever possible, it is desirable to use higher columns (around 2.0 m or greater), so that they are representative of the height of full-scale sedimentation tanks.

The test of the zone settling velocity can be done for various values of the initial concentration C_o, in order to allow the derivation of the parameters v_o and K from Equation 4.10.

Example 4.4

Determine the zone settling velocity of a suspension of activated sludge. The initial solids concentration of the mixed liquor in the column was equal to 2,900 mg/L. The following values of the interface height were measured as a function of time:

t (min)	0	3	6	9	12	15	18	21	24	27	30	45	60	90
H (m)	0.40	0.39	0.35	0.30	0.26	0.23	0.19	0.16	0.13	0.12	0.10	0.09	0.07	0.06

Solution:

a)   Plot the results on a graph

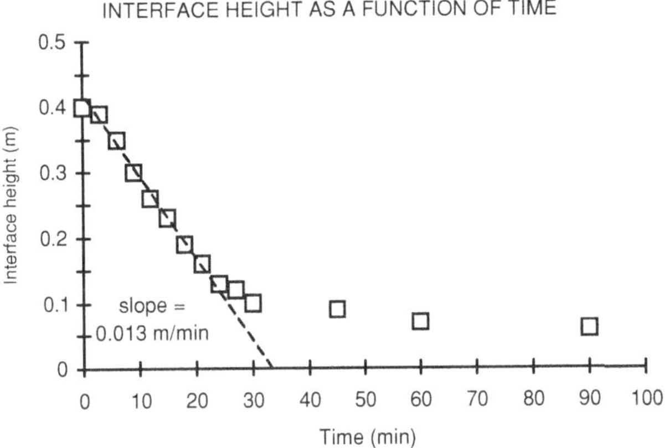

b)   Determine the slope of the straight line reach in the curve $H \times t$

Neglecting the initial point (t = 0 min) and the points starting from t = 21 min, which are not part of a straight line, the line of best fit is adjusted between the points t = 3 min and t = 18 min. The best line fit and its slope can be obtained graphically or through linear regression analysis. The values obtained graphically lead to the following value for the zone settling velocity:

$$v = 0.013 \text{ m/min} = 0.78 \text{ m/h}$$

To determine the values of the coefficients v_o and K from Equation 4.10, various settling tests should be undertaken for different values of the initial concentration C. In this way, distinct values of the pair v and C (zone settling sedimentation velocity $\times$ initial concentration) are obtained, allowing the determination of the coefficients v_o and K by graphical means or by a regression analysis.

Example 4.5

Determine the values of the coefficients v_o and K, based on the determination of the zone settling velocity v (according to the methodology of Example 4.4), for different values of the initial concentration C. The values obtained for v as a function of C were:

C (kg/m^3)	2.0	4.0	6.0	8.0	10.0	12.0
v (m/h)	2.03	0.55	0.13	0.04	0.01	0.00

Note: 1 kg/m^3 = 1000 g/m^3 = 1000 mg/l

Solution:

a)   Plot the curve of v × C

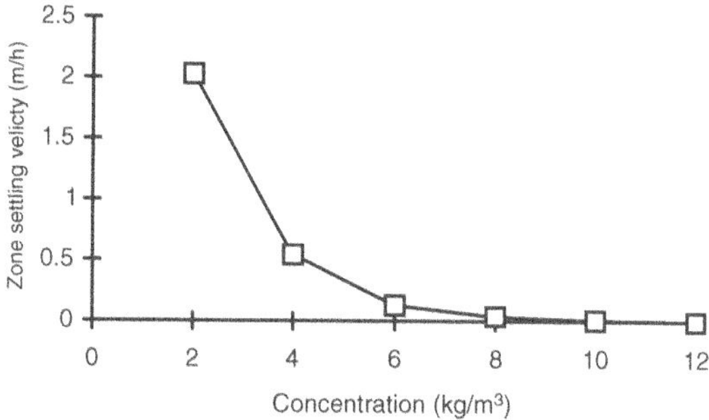

b)   Rearrange Equation 4.10 in a logarithmic form

Equation 4.10 is:

$$v = v_o . e^{-K.C}$$

Taking the natural logarithm of both sides of the equation:

$$\ln v = \ln v_o - K.C$$

The intercept of the line with the Y-axis is: $\ln v_o$
The slope of the line is: $-K$

c)   Plot the logarithmic form of Equation 4.10

The values of the logarithmic form of the equation are:

C (kg/m^3)	2.0	4.0	6.0	8.0	10.0	12.0
ln v	0.71	−0.60	−2.04	−3.21	−4.61	−

Example 4.5 (Continued)

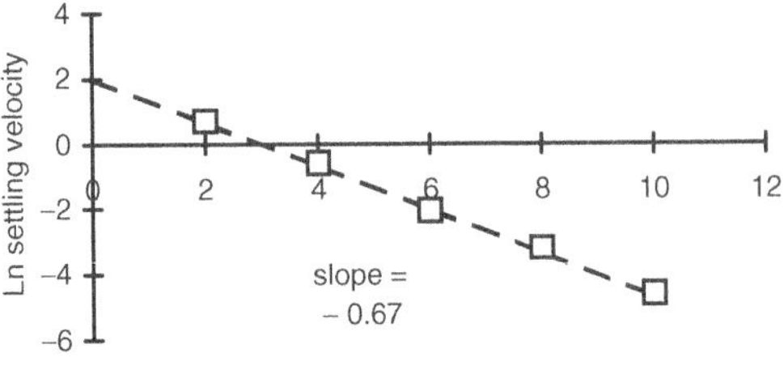

From the graph:

The intercept of the line with the Y-axis is: $\ln v_0 = 2.0 \rightarrow v_0 = e^{2.0} = 7.4$
The slope of the line is: $-K = -0.67 \rightarrow K = 0.67$

The coefficients of Equation 4.10 are, therefore:

$v_0 = 7.4$ m/h
$K = 0.67$ kg/m^3

These coefficients were, for didactic reasons, obtained graphically in the present example. However, it is preferable that these are obtained through regression analysis. The regression analysis can be linear (with the logarithmic transformation of Equation 4.10) or non-linear (with the original Equation 4.10).

4.5.4 Sludge volume index

The settleability of the sludge can be inferred through the settling curves, such as those presented in Section 4.5.3. However, frequently, in a wastewater treatment plant, only a simplified evaluation of the settleability is desired, aiming at using the data for the operational control of the plant. Under these conditions, the Sludge Volume Index (SVI) concept can be adopted.

The SVI is defined as the volume occupied by 1 g of sludge after settling for a period of 30 minutes. Hence, instead of determining the interface level at various time intervals, a single measurement at 30 minutes is made. The SVI is calculated through the following equation (see also Figure 4.19):

$$SVI = \frac{H_{30} \times 10^6}{H_0.SS} \tag{4.12}$$

where:
 SVI = Sludge Volume Index (mL/g)
 H_{30} = height of the interface after 30 minutes (m)

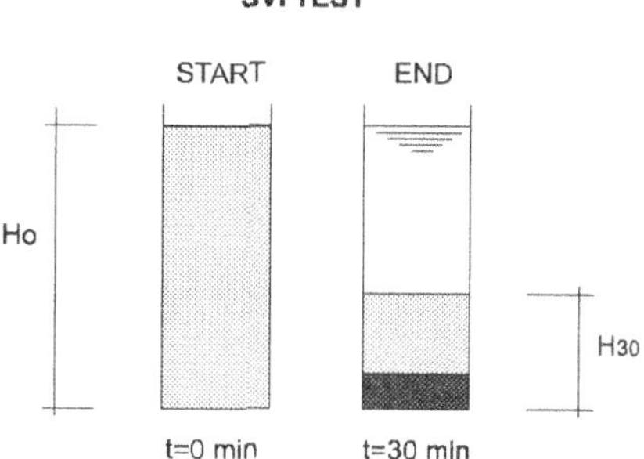

Figure 4.19.   Schematics of the SVI test

H_0 = height of the interface at time 0 (height of the water level in the settling cylinder) (m)

SS = suspended solids concentration in the sample (mg/L)

10^6 = conversion from mg to g, and from mL to L

Some standardisations are done in the SVI test, resulting in the following most common variants of the test:

- *Test without stirring during the settling period (SVI)*. The sample is left to settle without disturbances.
- *Test without stirring and with a dilution of the sample (DSVI)*. The original sample is diluted with the final effluent of the works in ratios of 2 (e.g. $^1/_2$, $^1/_4$, $^1/_8$, etc.). The DSVI is calculated using the diluted sample that leads to an interface height after 30 minutes (H_{30}) of less than 20% of the initial height (and as close as possible to 20%, with a tolerance of approximately 4%). The DSVI is calculated from Equation 4.12 and is multiplied by the reciprocal of the dilution ratio (e.g. multiplied by 4 if the dilution ratio was $^1/_4$).
- *Test with stirring during the settling period (SSVI)*. The stirring is mild and aims at reproducing the light stirring that occurs in a real scale settling tank. A thin vertical bar with a rotation of 1 or 2 rpm, situated between the centre and the periphery of the cylinder, causes the stirring in the cylinder.
- *Test with stirring and expression of the results in a standard concentration of 3.5 g/L (3500 mg/L) (SSVI$_{3.5}$)*. The advantage is that the results are expressed in a standard concentration (the interpretation of the other SVI tests is subject to the influence of the initial SS concentration). The concentration of 3.5 g/L is selected, because it represents an usual value of the mixed liquor suspended solids (MLSS) concentration in aeration tanks of the activated sludge process. This test is undertaken for different initial concentrations (obtained through dilutions and concentrations of the sample), and the results are interpolated for a concentration of 3.5 g/L. This test is the most representative and less subject to distortions.

Table 4.4. Approximate interpretation of values of the Sludge Volume Index (for activated sludge)

	Range of values for the Sludge Volume Index (mL/g)			
Settleability	SVI	DSVI	SSVI	$SSVI_{3.5}$
Excellent	0–50	0–45	0–50	0–40
Good	50–100	45–95	50–80	40–80
Fair	100–200	95–165	80–140	80–100
Poor	200–300	165–215	140–200	100–120
Very poor	>300	>215	>200	>120

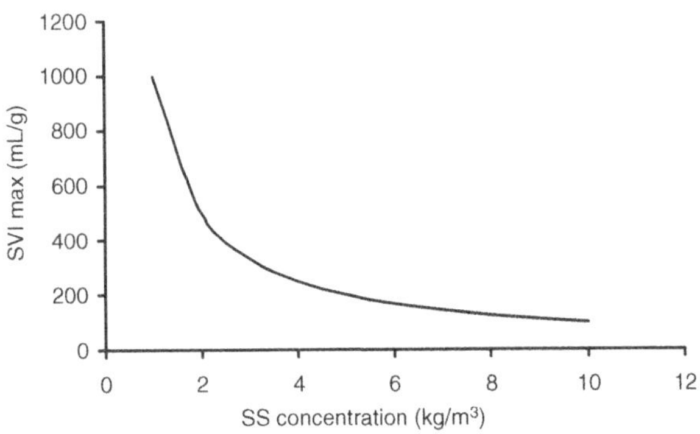

Figure 4.20. Maximum achievable SVI value (no sedimentation of the interface in the cylinder), as a function of the SS concentration.

The interpretation of the Sludge Volume Index is that, the larger the value, the lower the settleability of the sludge, that is, the sludge occupies a greater value in the secondary sedimentation tank. Besides this, the interpretation is also associated with the type of test. Typical approximate values are expressed in Table 4.4 (von Sperling, 1994; von Sperling and Fróes, 1999).

It should be emphasised that the traditional SVI test has a limitation because it is dependent on the initial solids concentration (denominator of Equation 4.12). For example, a sludge with a concentration of 1,000 mg/L that does not settle at all at the end of 30 minutes ($H_{30} = H_0$) will have a SVI of $10^6/1000 = 1.000$ mL/g. On the other hand, a sludge with a concentration of 10,000 mg/L that also does not settle after 30 minutes will have a SVI of $10^6/10,000 = 100$ mL/g. It is therefore clear the difficulty in the interpretation of the SVI results, because two sludges that do not settle at all have SVI values so different. Figure 4.20 shows the maximum SVI value (in which there is no settling in the cylinder) that can be obtained for sludges with different concentrations.

The values of DSVI and $SSVI_{3.5}$ are less susceptible to these influences, since they do not express the results in such varied concentrations. However, one should be always conscious of the fact that the SVI test and its variants only express the

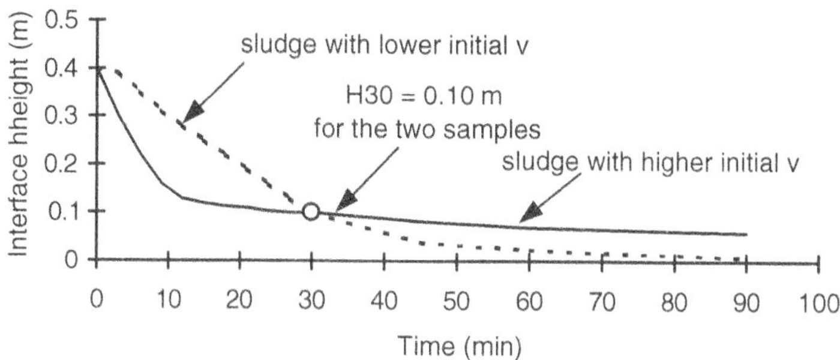

Figure 4.21.   Representation of two sludge samples with different settling velocities, but with the same SVI (adapted from Wanner, 1994)

sedimentation after a defined period (30 minutes) and do not give a direct indication of the settling velocity. Two sludges with the same SVI could have different settling velocities, as shown in Figure 4.21.

Example 4.6

Calculate the SVI (without stirring and dilution) of the activated sludge sample from Example 4.4 in which the following values were given or obtained:

- $H_0 = 0.40$ m
- $H_{30} = 0.10$ m
- SS = 2,900 mg/L

Solution:

From Equation 4.12:

$$SVI = \frac{H_{30} \times 10^6}{H_0.SS} = \frac{0.10 \times 10^6}{0.40 \times 2920} = 86 \text{ mL/g}$$

Based on the interpretation of Table 4.4, the settleability of this sludge can be considered good.

4.5.5 Determination of the limiting solids flux

As seen in Section 4.5.2, the total solids flux transmitted to the bottom of the secondary sedimentation tank is composed of the following components:

- *gravity flux* (G_g), caused by the gravitational sedimentation of the sludge;
- *underflow flux* (G_u), caused by the movement of the sludge resulting from the removal of the return sludge from the bottom of the sedimentation tank.

The respective formulas are:

Total flux:

$$\boxed{G_t = G_g + G_u} \tag{4.13}$$

Gravity flux:

$$\boxed{G_g = C.v_o.e^{-K.C}} \tag{4.14}$$

Underflow flux:

$$\boxed{G_u = C.\frac{Q_u}{A}} \tag{4.15}$$

where:

C = suspended solids concentration in the sludge (kg/m^3)

v_o = coefficient, expressing the zone settling velocity at a concentration $C = 0$ (m/h)

K = sedimentation coefficient (m^3/kg)

Q_u = sludge underflow (m^3/h)

A = surface area of the sedimentation tanks (m^2).

The limiting flux corresponds to the minimum at the curve G_t versus C. The minimum can be obtained, for a given value of Q_u/A, through the calculation of the limiting concentration C_L, such that the first derivative of the total flux equation G_t (Equation 4.13) is equal to zero, and the second derivative is greater than zero, to configure a minimum. The respective equations are:

- Limiting solids flux

$$G_L = v_o.C.e^{-K.C} + \frac{Q_u}{A} \tag{4.16}$$

- First derivative

$$G_L' = v_o.e^{-K.C}.(-K.C + 1) + \frac{Q_u}{A} = 0 \tag{4.17}$$

- Second derivative

$$G_L'' = v_o.e^{-K.C}.(C.K^2 - 2.K) > 0 \tag{4.18}$$

However, the determination of the limiting flux based on Equation 4.16 cannot be done directly. Because Equation 4.17 is not explicit in terms of C, it needs to be solved numerically by iteration (e.g. Newton–Raphson method) and the final result substituted again into Equation 4.16. Even though this solution can be obtained without problems using computer programs, this section presents the simpler and more didactic approach of the graphic solution, which can be also implemented in computers, using simple spreadsheets.

For given values of the coefficients v_o and K and of the sludge underflow velocity (Q_u/A), curves of the gravity flux, underflow flux and total flux can be composed graphically. Example 4.7 illustrates the methodology to be employed.

Example 4.7

Based on the data from Examples 4.4 and 4.5, compose the solids flux curves and determine the values of : (a) limiting solids flux, (b) limiting solids concentration and (c) solids concentration at the bottom sludge. Determine if the sedimentation tank is overloaded or underloaded.

Data given in Examples 4.4 and 4.5:

* $v_o = 7.4$ m/h
* $K = 0.67$ m^3/kg
* MLSS: $C_o = 2,900$ g/m$^3 = 2.9$ kg/m^3

Additional data:

* total surface area of the secondary sedimentation tanks: $A = 500$ m^2
* influent wastewater flow to the works: $Q_i = 350$ m^3/h
* return sludge flow ($\approx$ underflow): $Q_u = 200$ m^3/h

Solution:

a) Calculate the fluxes for different values of solids concentration

The gravity flux, underflow flux and total flux are calculated below, for values of C varying from 0 to 20 kg/m^3.

C (kg/m^3)	v (m/h)	G_g (kg/m^2 h)	G_u (kg/m^2.h)	G_t (kg/m^2.h)
0.0	7.40	0.00	0.00	0.00
0.5	5.29	2.65	0.20	2.85
1.0	3.79	3.79	0.40	4.19
1.5	2.71	4.06	0.60	4.66
2.0	1.94	3.88	0.80	4.68
2.5	1.39	3.47	1.00	4.47
3.0	0.99	2.97	1.20	4.17
3.5	0.71	2.48	1.40	3.88
4.0	0.51	2.03	1.60	3.63
4.5	0.36	1.63	1.80	3.43
5.0	0.26	1.30	2.00	3.30
5.5	0.19	1.02	2.20	3.22
6.0	0.13	0.80	2.40	3.20
6.5	0.10	0.62	2.60	3.22
7.0	0.07	0.48	2.80	3.28
7.5	0.05	0.36	3.00	3.36
8.0	0.03	0.28	3.20	3.48
8.5	0.02	0.21	3.40	3.61

Example 4.7 (Continued)

C (kg/m^3)	v (m/h)	G_g $(kg/m^2.h)$	G_u $(kg/m^2.h)$	G_t $(kg/m^2.h)$
9.0	0.02	0.16	3.60	3.76
9.5	0.01	0.12	3.80	3.92
10.0	0.01	0.09	4.00	4.09
10.5	0.01	0.07	4.20	4.27
11.0	0.00	0.05	4.40	4.45
11.5	0.00	0.04	4.60	4.64
12.0	0.00	0.03	4.80	4.83
12.5	0.00	0.02	5.00	5.02
13.0	0.00	0.02	5.20	5.22
13.5	0.00	0.01	5.40	5.41
14.0	0.00	0.01	5.60	5.61
14.5	0.00	0.01	5.80	5.81
15.0	0.00	0.00	6.00	6.00
15.5	0.00	0.00	6.20	6.20
16.0	0.00	0.00	6.40	6.40
16.5	0.00	0.00	6.60	6.60
17.0	0.00	0.00	6.80	6.80
17.5	0.00	0.00	7.00	7.00
18.0	0.00	0.00	7.20	7.20
18.5	0.00	0.00	7.40	7.40
19.0	0.00	0.00	7.60	7.60
19.5	0.00	0.00	7.80	7.80
20.0	0.00	0.00	8.00	8.00

where:

- C = *suspended solids concentration, varying from 0 to 20 kg/m³*
- v = *zone settling velocity (m/h)*
  Given by Equation 4.10: $v = v_0.e^{-K.C}$
  In example 4.4, the values of the coefficients v_0 and K were determined (v_0 = 7.4 m/h and K = 0.67 kg/m³)
- G_g = *gravity solids flux (kg/m².h)*
  Given by Equation 4.14: $G_g = C.v_0.e^{-K.C}$
- G_u = *underflow flux (kg/m².h)*
  Given by Equation 4.15: $G_u = C.(Q_u/A)$
  The value of Q_u/A is calculated based on the data given in the problem:
  $Q_u/A = (200 \text{ m}^3/h) / (500 \text{ m}^2) = 0.4$ m/h
- G_t = *total solids flux (kg/m².h)*
  Given by Equation 4.13: $G_t = G_g + G_u$

b)   Plot the gravity flux (G_g)

As shown in Figure 4.16, one method for determining the limiting flux is through the graph of the gravity flux, presented below. The data necessary for the composition of the graph are taken from the previous table.

Example 4.7 (Continued)

Gravity flux

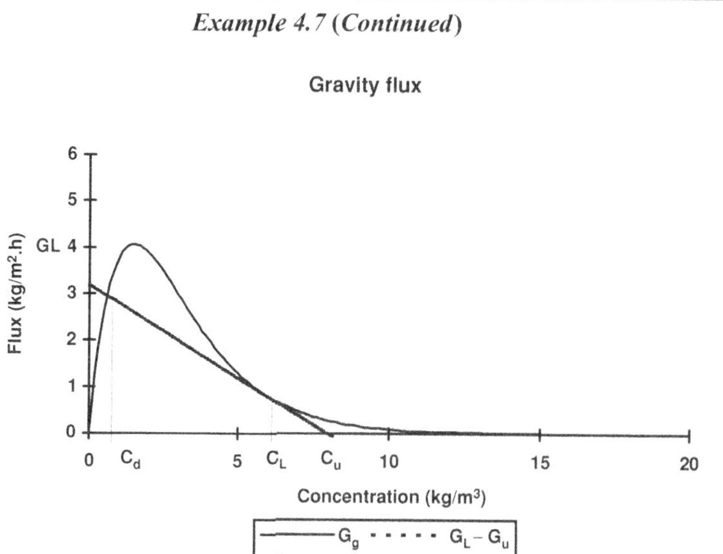

The curve of the flux G_g is taken directly from the values in the previous table. The dashed line (characterised as $G_L - G_u$) has a slope of Q_u/A, and is located as tangent to the curve G_g. This line presents four important points:

- intercept with the Y-axis: *limiting flux* G_L (obtained on the Y-axis)
- intercept with the curve G_g: *concentration of the diluted layer* C_d (obtained on the X-axis)
- tangent to the curve G_g: *limiting concentration* C_L (obtained on the X-axis)
- intercept with the X-axis: *underflow sludge concentration* C_u (obtained on the X-axis)

The values obtained graphically are:

- *limiting flux* G_L: 3.2 kg/m².h
- *concentration of the diluted layer* C_d: 0.6 kg/m³
- *limiting concentration* C_L: 6.0 kg/m³
- *underflow sludge concentration* C_u: 8.0 kg/m³

c)   Plot the total flux (G_t)

An alternative form of determining the limiting flux is through the total flux graph, discussed in Figure 4.16. The necessary values for the composition of the graph are taken directly from the previous table (G_g, G_u and G_t).

Example 4.7 (Continued)

Total flux

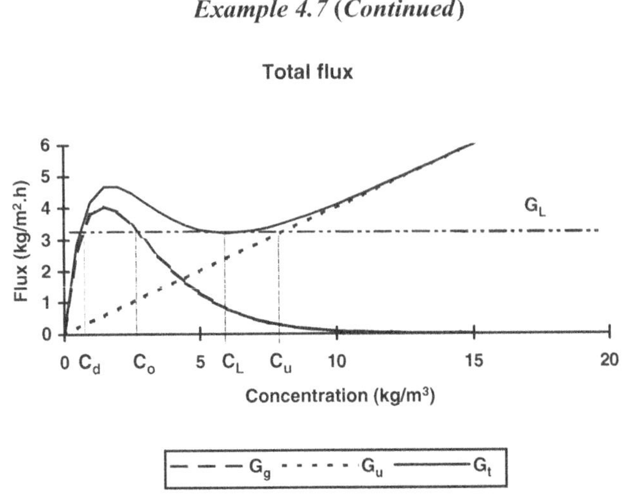

   The limiting flux corresponds to the minimum of the total flux curve G_g. The other parameters (C_d, C_L and C_u) can be obtained based on the intercepts of the tangent to this point of minimum (parallel with the X-axis) with the curve G_g, as seen in the graph. Naturally, the values obtained are the same as those already obtained in the previous graph presented in item b.

d)   Interpretation of the flux curves

The mixed liquor enters the sedimentation tank with a concentration of $C_o = 2.9$ kg/m³ (MLSS). While the sludge thickens and its concentration increases, the capacity to transmit solids to the bottom of the tank decreases (see the curve of G_t in the above graph). This occurs because in this range the increase of C brings about a reduction in the settling velocity v (see Equation 4.10). The capacity to transfer these solids to the bottom decreases until the concentration of 6 kg/m³ (limiting concentration C_L) is reached. In these conditions there is the maximum value of the flux that can be transferred to the bottom, that is, the limiting flux ($G_L = 3.2$ kg/m².h) (see the point of minimum on the curve G_t). Subsequently, the sludge concentration increases until it reaches the concentration of the underflow sludge at the bottom ($C_u = 8.0$ kg/m³). In this range, the transmission capacity of the solids to the bottom starts to increase again (see curve G_t above), due to the contribution of C in the gravity flux G_g (see Equation 4.14). In spite of this increase of the flux transmitted to the bottom, the sedimentation tank is previously limited by its limiting flux capacity. If a solids flux greater than the limiting flux is applied in the tank, the applied flux will not be able to be totally transmitted to the bottom of the tank, because it is greater than the limiting flux. In these conditions, only the limiting flux is transferred, and the flux

---

Example 4.7 (Continued)

in excess generates an expansion in the volume occupied by the sludge (rising of the sludge blanket level).

e)   Evaluation of the loading conditions

To determine if the sedimentation tank is overloaded or underloaded, the applied and limiting fluxes must be compared. As seen, the limiting flux is $G_L = 3.2$ kg/m^2.h. The applied flux G_a is given by Equation 4.11:

$$G_a = \frac{Q_i + Q_r}{A}.C_o = \frac{(350 + 200)}{500} \times 2.9 = 3.2\, \text{kg/m}^2.\text{h}$$

It can be observed that the applied flux is equal to the limiting flux, or $G_a = G_L$. In these conditions, the sedimentation tank is in equilibrium, and the level of the sludge blanket remains constant.

Example 4.8

The influent flow to the plant analysed in Example 4.7 suffers an increase from 350 m^3/h to 450 m^3/h. Analyse the impact of this increase and propose a control measure.

Solution:

a)   Evaluation of the loading conditions

Since the settleability of the sludge (coefficients v_o and K) and the underflow (Q_u) were not altered, the limiting flux remains the same ($G_L = 3.2$ kg/m^2.h). However, the applied flux increased due to the increase of Q_i. The new applied flux becomes:

$$G_a = \frac{Q_i + Q_r}{A}.C_o = \frac{(450 + 200)}{500} \times 2.9 = 3.8\, \text{kg/m}^2.\text{h}$$

In these conditions, $G_a > G_L$. The sedimentation tank cannot transmit to the bottom the total applied flux (3.8 kg/m^2.h), but only the limiting flux (3.2 kg/m^2.h). The exceeding flux ($3.8 - 3.2 = 0.6$ kg/m^2.h) will not be able to go to the bottom and will cause an increase in the volume of the sludge layer.

If no operational measures are taken, the persistence of this excess flux will cause a continuous expansion of the sludge blanket, until the solids start to leave with the final effluent. After that, a new equilibrium situation will be reached, with the excess flux leaving with the final effluent, and causing a greater deterioration in its quality. The solids load that leaves with the effluent

Example 4.8 (Continued)

will be $0.6 \text{ kg/m}^2.\text{h} \times 500 \text{ m}^2 = 300 \text{ kg/h}$. This load represents a concentration in the final effluent of $(300 \text{ kg/h}) \div (450 \text{ m}^3/\text{h}) = 0.67 \text{ kg/m}^3 = 670 \text{ mg/L}$. This concentration is obviously unacceptable for the final effluent of a wastewater treatment plant.

b) Control measures

To avoid the expansion of the sludge blanket, the following operational measures can be taken:

- increase of the underflow rate
- reduction of the MLSS concentration

Increase of Q_u. If the underflow rate is increased, the limiting and applied fluxes will increase. However, there is a value of Q_u that permits both fluxes to be the same. Adopting the methodology of Example 4.7 for different values of Q_u, the value that leads to this new equilibrium condition can be obtained. In this case, the value found was $Q_u = 315 \text{ m}^3/\text{h}$. The limiting and applied fluxes increase to $4.44 \text{ kg/m}^2.\text{h}$. Since both are equal, there will be no expansion of the sludge blanket. Hence, the increase of the underflow rate (that is, also of the return sludge flow) is an effective measure for the control of a secondary sedimentation tank subjected to an increase in the influent load.

Reduction of MLSS. One way of decreasing the applied flux is by reducing the influent concentration to the sedimentation tank, that is, MLSS. In this case, the applied flux is reduced and the limiting flux remains the same. If MLSS is reduced from 2.9 kg/m^3 to 2.45 kg/m^3, the new applied flux will be $3.2 \text{ kg/m}^2.\text{h}$, equal to the limiting flux (determined in Example 4.7). The reduction in the MLSS concentration is accomplished by the increase in the wastage flow of the excess (surplus) sludge. However, the capacity of the sludge treatment system to receive this increased load of the excess sludge needs to be verified.

Although these analyses seem laborious, it should be remembered that these calculations are easily implemented in a computer program or spreadsheet.

It should be also remembered that these analyses are applicable to steady-state conditions. However, the principles of the limiting solids flux theory are also applicable to dynamic models.

5

Aeration

5.1 INTRODUCTION

Aeration is a unit operation of fundamental importance in a large number of aerobic wastewater treatment processes. When a liquid is deficient in a gas (oxygen, in this case), there is a natural tendency of the gas to pass from the gas phase, where it is present in sufficient concentrations, to the liquid phase, where it is deficient. Oxygen is a gas that dissolves poorly in the liquid medium. For this reason, in various wastewater treatment systems it is necessary to accelerate the natural process, in such a way that the oxygen supply may occur at a higher rate, compatible with the biomass utilisation rate. Among the wastewater treatment processes that use artificial aeration are aerated lagoons, activated sludge and its variants, aerated biofilters and other more specific processes. In terms of sludge treatment, aerobic digesters also use artificial aeration.

There are two main forms of producing artificial aeration:

- introduce air or oxygen into the liquid (*diffused air aeration*)
- cause a large turbulence, exposing the liquid, in the form of droplets, into the air, and also permitting the entrance of atmospheric air into the liquid medium (*surface or mechanical aeration*)

Within these two types there are variants, which are described in Sections 5.7 and 5.8. Figure 5.1 presents schematically the principles of aeration by diffused air and mechanical aeration.

In the end of this chapter, another way of transferring oxygen to the liquid medium is described: gravity aeration (steps, weirs, cascades).

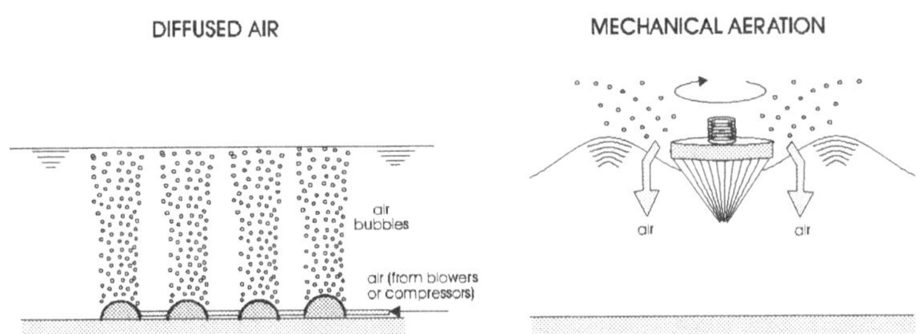

Figure 5.1.   Schematic representation of diffused air and mechanical aeration systems.

5.2 FUNDAMENTALS OF GAS TRANSFER

5.2.1 Saturation concentration of a gas

When a liquid is exposed to a gas, there is a continuous exchange of molecules from the liquid phase to the gas phase and vice versa. As soon as the solubility concentration in the liquid phase is reached, both fluxes become equal in magnitude, in such a way that no overall change in the gas concentrations in both phases occurs (Figure 5.2). This dynamic equilibrium is associated with the **saturation concentration** of the gas in the liquid phase.

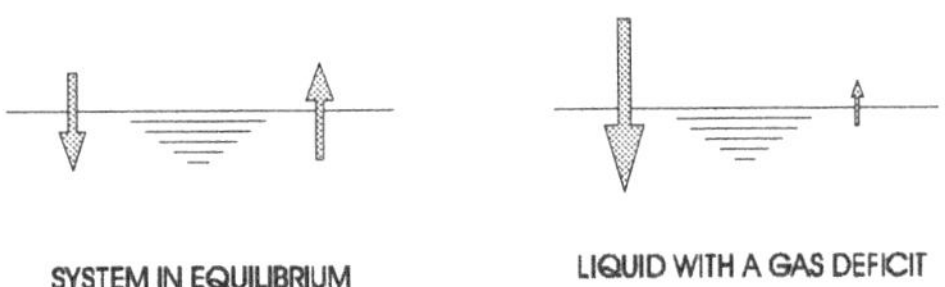

Figure 5.2.   Gas exchanges in (a) a system in equilibrium and (b) a system with a deficit in the dissolved gas

In equilibrium conditions, the velocities of absorption (v_g) and release (v_l) of the gas are equal, that is:

$$v_g = v_l \qquad (5.1)$$

The saturation concentrations in the two phases are proportional to these velocities:

$$k_g.C_g = k_l.C_s \qquad (5.2)$$

where:
k_g and k_l = constants of proportionality
C_g = concentration of the gas in the gas phase (mg/L)
C_s = concentration of the gas in the liquid phase (mg/L)

Hence,

$$C_s = \frac{k_g}{k_l}.C_g \qquad (5.3)$$

Making $k_g/k_l = k_D$:

$$\boxed{C_s = k_D.C_g} \qquad (5.4)$$

Therefore, the saturation concentration is directly proportional to the concentration in the gas phase (Henry's Law). The coefficient k_D is called the *distribution coefficient*, and depends on the nature of the gas and the liquid, and the temperature. For the diffusion of oxygen in water, k_D values are (Pöpel, 1979):

Table 5.1. Values of the distribution coefficient k_D for oxygen

Temperature (°C)	k_D
0	0.0493
10	0.0398
20	0.0337
30	0.0296

Therefore, it can be seen that, the greater the temperature, the lower the solubility of the gas in the liquid medium. The larger agitation of the molecules in the water contributes to the transfer of the gas to the gas phase.

The concentration C_g can be obtained from the universal gas law:

$$pV = nRT \qquad (5.5)$$

where:
   p = partial pressure of the gas in the gas phase (Pa)
   V = volume occupied in the gas phase (m^3)
   n = number of moles of gas in the volume V (mol/m^3)
   R = universal constant (8.3143 J/K.mol)
   T = temperature (K)

Developing Equation 5.5 and introducing it into Equation 5.4 and, at the same tine correcting it for the water vapour pressure, the equation that establishes the saturation concentration of a gas in water as a function of the temperature and pressure is obtained:

$$C_s = k_D.d_v.(P_a - P_v).\frac{MW}{R.T} \qquad (5.6)$$

where:
   d_v = volumetric distribution of oxygen in atmospheric air (0.21 or 21% of the
       air, in volume, is represented by oxygen)

MW = molecular weight of oxygen (32 g/mol)

P_a = atmospheric pressure (101,325 Pa in standard temperature and pressure)

P_v = water vapour pressure (see Table 5.2)

Table 5.2. Water vapour pressure values (P_v)

Temperature (°C)	P_v (Pa)
0	611
10	1,230
20	2,330
30	4,240

Example 5.1

Calculate the saturation concentration of oxygen in pure water, in the following conditions:

- temperature = 20 °C
- atmospheric pressure at sea level

Solution:

Based on the values presented above:

$k_D = 0.337$

$d_v = 0.21$

$P_a = 101,325$ Pa

$P_v = 2,330$ Pa

MW = 32 g/mol

R = 8.3143 J/K.mol

T = 293 K(= 20 °C)

Using Equation 5.6:

$$C_s = k_D.d_v.(P_a - P_v).\frac{MW}{R.T}$$

$$= 0.0337 \times 0.21 \times (101,325 - 2,330).\frac{32}{8.3143 \times 293} = 9.2 \text{ mg/L}$$

In order to avoid this sequence of laborious calculations, there are some empirical formulas available in the literature (the majority based on regression analysis) that give directly the value of C_s(mg/L) as a function, for example, of the temperature T (°C). A formula frequently used is:

$$C_s = 14.652 - 4.1022 \times 10^{-1}.T + 7.9910 \times 10^{-3}.T^2 - 7.7774 \times 10^{-5}.T^3$$

$$(5.7)$$

Altitude exerts an influence on the solubility of a gas, because it is inversely proportional to the atmospheric pressure. The greater the altitude, the lower the atmospheric pressure and the lower the pressure for the gas to be dissolved in the water. This influence can be computed by the following relation (Qasim, 1985):

$$f_H = \frac{C'_s}{C_s} = \left(1 - \frac{H}{9450}\right)$$

(5.8)

where:

f_H = correction factor for the DO saturation concentration by the altitude (−)
C'_s = saturation concentration at the altitude H (mg/L)
H = altitude (m)

The salinity also affects the solubility of the oxygen. The influence of dissolved salts can be computed by the following empirical formula (Pöpel, 1979):

$$\gamma = 1 - 9 \times 10^{-6}.C_{sal}$$

(5.9)

where:

γ = solubility reduction factor ($\gamma = 1$ for pure water)
C_{sal} = concentration of dissolved salts (mg Cl$^-$/l)

5.2.2 Gas transfer mechanisms

5.2.2.1 Molecular diffusion

There are two basic mechanisms for the transfer of oxygen from the gas phase to the liquid phase:

- *molecular diffusion*
- *turbulent diffusion*

Molecular diffusion can be understood as the tendency of any substance to spread itself uniformly in the space available.

For a water body of unlimited depth, exposed to the gas phase through a surface A, the mass transfer rate dM/dt due to the diffusion of the gas molecules in the liquid phase is defined by Fick's law (Pöpel, 1979):

$$\frac{dM}{dt} = -D.A.\frac{\partial C}{\partial x}$$

(5.10)

where:

D = coefficient of molecular diffusion (m^2/s)
A = surface area (m^2)

$$x = \text{distance from the interface (m)}$$
$$\partial C/\partial x = \text{concentration gradient } (g/m^3.m)$$

It is important to note that, for a certain gas, only the concentration gradient determines the diffusion rate per unit area. The negative sign indicates that the direction of diffusion is opposite to the positive concentration gradient.

For oxygen, the values of the diffusion coefficient are presented in Table 5.3.

Table 5.3. Values of the diffusion coefficient D

Temperature (°C)	D (10^{-9} m^2/s)
10	1.39
20	1.80
30	2.42

Two theories frequently used to explain the gas transfer mechanism are (Pöpel, 1979):

- *Two-film theory*. In the gas–liquid interface there are two films, a gas film and a liquid film. The gas is absorbed and transported by molecular diffusion and mixing (convection) by the gas film and subsequently by the liquid film. The films are considered as stagnant and with a fixed thickness. The two-film theory is simpler but provides a good answer in most cases (Metcalf & Eddy, 1991).
- *Penetration theory*. The penetration theory does not assume stagnant films, but fluid elements that are momentarily exposed to the gas phase in the liquid interface. During this exposure time the gas diffuses in the fluid elements *penetrating* the liquid. Differently from the two-film theory, the penetration process is described by an unsteady diffusion. The exposure time is considered very short (< 0.1 s) for steady diffusion conditions to prevail. The penetration theory is more soundly theoretically based.

According with the penetration theory, the following formulas for the gas transfer can be obtained:

- Absorption rate of the gas:

$$\frac{dM}{dt} = A.(C_s - C_o).\sqrt{\frac{D}{\pi.t}} \qquad (5.11)$$

- Penetration depth of the gas:

$$x_p = \sqrt{\pi.D.t} \qquad (5.12)$$

where:
  M = mass of the gas absorbed through the area A during the time t (g)
  A = interfacial exposure area (m^2)
  t = exposure time (s)

C_o = initial concentration of the gas in the bulk of the liquid mass (g/m^3)

x_p = penetration depth of the gas in the liquid mass (m)

Example 5.2

A tank under quiescent conditions, completely deprived of oxygen, with a temperature of 20 °C, at sea level, is exposed to air.

- What is the absorption rate of oxygen?
- What is the penetration depth of oxygen?

Solution:

a)   Oxygen absorption rate

$C_o = 0.0 \ g/m^3$
$C_s = 9.2 \ g/m^3$
$D = 1.8 \times 10^{-9} \ m^2/s$ (see Table 5.3)

$$\frac{dM}{dt.A} = (C_s - C_0).\sqrt{\frac{D}{\pi.t}} = (9.2 - 0.0).\sqrt{\frac{1.8 \times 10^{-9}}{\pi.t}}$$

After 1 second (t = 1 s)   = $220 \times 10^{-6} \ g/m^2.s$
      1 minute (t = 60 s) = $28 \times 10^{-6} \ g/m^2.s$
      1 hour (t = 3,600 s) = $3.7 \times 10^{-6} \ g/m^2.s$
      1 day (t = 86,400 s) = $0.75 \times 10^{-6} \ g/m^2.s$

b)   Penetration depth of the oxygen

$$x_p = \sqrt{\pi.D.t} = \sqrt{\pi.1.8 \times 10^{-9}.t}$$

After 1 second (t = 1 s)   = $0.075 \times 10^{-3} \ m$
      1 minute (t = 60 s) = $0.582 \times 10^{-3} \ m$
      1 hour (t = 3,600 s) = $4.51 \times 10^{-3} \ m$
      1 day (t = 86,400 s) = $22.1 \times 10^{-3} \ m$

The objective of the present example is to emphasise the fact that the transfer of oxygen by molecule diffusion is extremely slow. In wastewater treatment, the high oxygen demand cannot by supplied simply by molecular diffusion.

5.2.2.2 Turbulent diffusion

In sewage treatment with artificial aeration, the main gas transfer mechanism occurs through the *creation* and *renewal* of the interfaces.

The turbulent flux generated by artificial aeration consists of a complex secondary movement that surpasses the primary movement of the liquid mass. The turbulence is characterised by oscillations and eddies that transport fluid particles from one layer to another, with variable velocities. The turbulent movement, which is erratic in direction, magnitude and time, can be defined only probabilistically (O'Connor and Dobbins, 1958).

As mentioned, gas transfer by turbulent diffusion is much higher than by molecular diffusion. The basic structure of the gas transfer formulation can be maintained, with adaptations only in the sense of simplifying its presentation. The transfer coefficient incorporates other constants, as shown in Sections 5.3 and 5.4.

5.3 KINETICS OF AERATION

The disadvantage of the mentioned formulas is that the diffusion coefficient D, the exposure time to the interfacial area t_c and the interfacial area A must be known, in order to allow the estimation of the gas transfer rate. Consequently, it is necessary to adopt a more practical approach, as discussed below.

Under steady-state conditions, the diffusion coefficient D and the exposure time t_c can be assumed as constants, resulting in a constant gas transfer coefficient (K_L). Besides this, the surface area A and the specific area a ($=$ area/volume $=$ A/V) may also be assumed as constants. Under these conditions, a constant $K_L a$ can be defined:

$$K_L\,a = 2.\sqrt{\frac{D}{\pi.t_c}}.\frac{A}{V} \qquad (5.13)$$

where:
$K_L a$ = overall oxygen mass transfer coefficient (s^{-1})

Thus, the mass of oxygen transferred per unit time and volume can be expressed in the following simplified form, through a simple rearrangement of Equation 5.13:

$$\frac{m}{V} = K_L a.(C_s - C) \qquad (5.14)$$

or:

$$\boxed{\frac{dC}{dt} = K_L a.(C_s - C)} \qquad (5.15)$$

where:
dC/dt = rate of change of the oxygen concentration $(g/m^3.s)$
C = concentration at any time t (g/m^3)

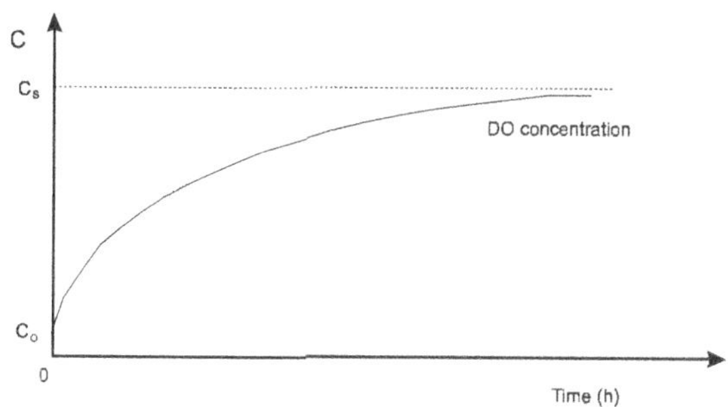

Figure 5.3.   Temporal progression of the DO concentration during aeration (*without oxygen consumption*)

Through Equation 5.15, it can be seen that *the lower the oxygen concentration (C), or the higher the oxygen deficit (C_s − C), the greater is the oxygen transfer rate (dC/dt).*

Integrating Equation 5.15 between the limits of t = 0 to t = t and C = 0 to C = C leads to (ASCE, 1990):

$$\frac{C_s - C}{C_s - C_o} = e^{-K_L a.(t-t_o)} \tag{5.16}$$

If there is *no oxygen consumption* in the liquid medium under aeration (example: clean water), the concentration C increases according to a decreasing exponential rate (observe the negative sign on the exponent of e). The concentration tends asymptotically to the steady-state value, that is, the saturation concentration C_s. The formula of this trajectory is obtained through the rearrangement of Equation 5.16, being expressed in Equation 5.17 and represented in Figure 5.3.

$$C = C_s - (C_s - C_o).e^{-K_L a.(t-t_o)} \tag{5.17}$$

In case there is oxygen consumption in the liquid medium, which occurs in aeration tanks and aerated lagoons, the highest value that can be reached by the oxygen concentration is lower than the saturation value. Designating the oxygen consumption rate by **r** ($g/m^3.s$) and the maximum value to be reached by C as C_∞, the equation of the trajectory of DO in a reactor *with oxygen consumption* is (see Figure 5.4):

$$C = C_\infty - (C_\infty - C_o).e^{-K_L a.(t-t_o)} \tag{5.18}$$

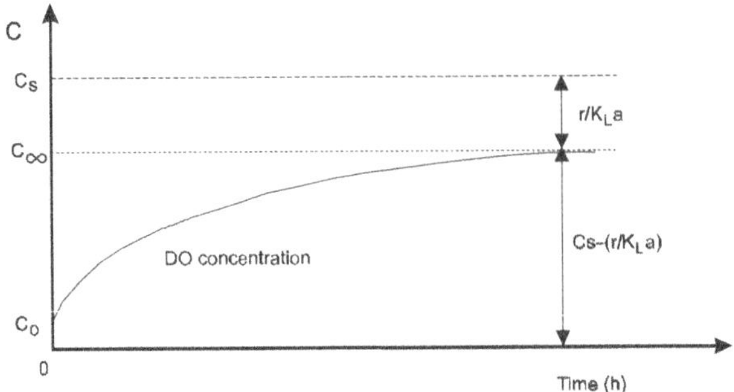

Figure 5.4. Temporal progression of the DO concentration during aeration (*with* oxygen consumption)

In the steady state, in a reactor under operation, the oxygen consumption rate (or oxygen utilisation rate) by the bacteria is equal to the oxygen production rate by the aeration system. Hence:

$$O_2 \text{ production rate} = O_2 \text{ consumption rate}$$

$$K_L a.(C_s - C) = r \tag{5.19}$$

or:

$$\boxed{C = C_s - \frac{r}{K_L a}} \tag{5.20}$$

The value of C obtained in Equation 5.20 corresponds to C_∞, presented in Figure 5.4.

The experimental determination of the coefficient $K_L a$ is discussed in Section 5.9.

5.4 FACTORS OF INFLUENCE IN OXYGEN TRANSFER

The oxygen transfer rate of the aeration equipment to be installed in a wastewater treatment plant is frequently determined in different conditions under which it will operate (*operating conditions*). Therefore, it is important to be able to quantify the factors that influence the oxygen transfer rate, to allow the estimation of the transfer rate under operating conditions, based on results obtained in tests undertaken under standardised conditions.

The factors of major influence on the oxygen transfer rate are:

- temperature
- atmospheric pressure (altitude)
- dissolved oxygen concentration

- characteristics of the wastewater
- characteristics of the aerator and the geometry of the reactor

a) Temperature

The influence of temperature occurs according to two apparently opposite directions:

- *Influence on the saturation concentration C_s*. The increase of the temperature causes a **reduction** in the saturation concentration C_s, which implies a **reduction** in the transfer rate dC/dt.
- *Influence on the mass transfer coefficient $K_L a$*. The increase in the temperature causes an **increase** in the coefficient $K_L a$, which implies an **increase** in the transfer rate dC/dt.

The influence on the saturation concentration was discussed in Section 5.2.1. The influence on $K_L a$ can be expressed by:

$$K_L a_{(T)} = K_L a_{(20\,°C)}.\theta^{(T-20)} \tag{5.21}$$

where:

$K_L a_{(T)}$ = coefficient $K_L a$ at any temperature T (s^{-1})
$K_L a_{(20)}$ = coefficient $K_L a$ at a temperature of 20 °C (s^{-1})
θ = temperature coefficient. Usually adopted as 1.024.

b) Atmospheric pressure (altitude)

The influence of the altitude is manifested in the oxygen saturation concentration (the greater the altitude, the lower the atmospheric pressure and, therefore, the lower the saturation concentration). The correction factor for the altitude was discussed in Section 5.2.1.

c) Dissolved oxygen concentration

Under steady-state conditions, the greater the dissolved oxygen concentration (C) maintained in the reactor, the lower the value of $C_s - C$, that is, the lower is the oxygen transfer rate (see Equation 5.15). For example, in activated sludge systems, the DO concentration maintained in the reactor is usually in the range of 1.0 to 2.0 mg/L.

The correction of the influence of the DO concentration is obtained by:

$$\text{Correction factor for the DO concentration} = \frac{C_{sw} - C_L}{C_s(20\,°C)} \tag{5.22}$$

where:

C_{sw} = saturation concentration of DO in the liquid in the reactor (mg/L) (see item d below)
C_L = DO concentration maintained in the liquid in the reactor (mg/L)
$C_s(20\,°C)$ = saturation concentration of DO in clean water at 20 °C (mg/L)

d) Wastewater and reactor characteristics

The specific characteristics of the wastewater being treated and the configuration of the reactor, which are different from the test conditions in which the oxygen transfer is measured, also exert an influence on the actual transfer rate in the field, under operating conditions. This influence occurs in two ways:

- influence on the oxygen saturation concentration in the liquid in the reactor (C_{sw})
- influence on the oxygen transfer coefficient (K_La)

Influence on C_{sw}. The presence of salts, particulate matter and detergents affect the saturation concentration of the liquid in the reactor. This influence can be quantified through the following correction factor:

$$\beta = \frac{C_{sw}(\text{wastewater})}{C_s(\text{clean water})} \tag{5.23}$$

The values of β vary from 0.70 to 0.98, but the value of 0.95 is frequently adopted (Metcalf & Eddy, 1991).

Influence on K_La. The oxygen transfer coefficient is influenced by the characteristics of the wastewater as well as the geometry of the reactor and mixing level. The correction factor is:

$$\alpha = \frac{K_La\,(\text{wastewater})}{K_La\,(\text{clean water})} \tag{5.24}$$

Typical values of α vary from 0.6 to 1.2 for mechanical aeration and from 0.4 to 0.8 for diffused air aeration (Metcalf & Eddy, 1991).

5.5 OXYGEN TRANSFER RATE IN THE FIELD AND UNDER STANDARD CONDITIONS

The oxygen transfer rate will vary from place to place, for the same equipment, due to the simultaneous interaction of the various factors covered in Section 5.4. Therefore, it is important that the transfer rate may be expressed under standard conditions, in order to allow a uniform presentation of the values. Therefore, there are the following two ways of expressing the oxygen transfer rate (oxygenation capacity):

Standard conditions	*Operating conditions (field)*
• clean water	• wastewater
• liquid temperature = 20 °C	• real temperature of the liquid
• altitude = 0 m (sea level)	• real altitude of the plant
• aeration system installed in a test tank	• aeration system installed in the actual reactor

The conversion of one form to the other is done with the correction factors presented in Section 5.4. Incorporating all these factors, the general conversion equation is obtained:

$$OTR_{standard} = \frac{OTR_{field}}{\dfrac{\beta.f_H.C_s - C_L}{C_s(20\,^\circ C)}.\alpha.\theta^{T-20}}$$

(5.25)

where:

$OTR_{standard}$ = Standard Oxygen Transfer Rate – **SOTR** (kgO_2/h)

OTR_{field} = Oxygen Transfer Rate in the field, under operating conditions (kgO_2/h)

C_s = oxygen saturation concentration in clean water, at the operating temperature in the field (g/m^3)

C_L = average concentration of oxygen maintained in the reactor (g/m^3)

$C_s(20\,^\circ C)$ = saturation concentration of oxygen in clean water, under standard conditions (g/m^3)

f_H = correction factor C_s for the altitude ($= 1 - $ altitude/9450) (see Equation 5.8)

β = see comments for Equation 5.23

α = see comments for Equation 5.24

θ = see comments for Equation 5.21

T = liquid temperature ($^\circ C$)

The relation between the standard oxygen transfer rate (SOTR or $OTR_{standard}$) and the oxygen transfer coefficient (K_La) can be obtained through rearrangement of Equation 5.15, in which $C = 0$ (standard conditions), leading to:

$$OTR_{standard} = \frac{K_La.C_{s(20\,^\circ C)}.V}{1000}$$

(5.26)

Due to the various influencing factors, the OTR_{field} is lower than the $OTR_{standard}$. Thus, in the designs, usually OTR_{field} is estimated as a function of the oxygen requirements and subsequently the $OTR_{standard}$ is calculated using Equation 5.25.

Example 5.3

In a wastewater treatment plant the supply of 100 kgO_2/h is necessary under operating conditions, using a mechanical aeration system. Determine the Standard Oxygen Transfer Rate knowing that:

- Liquid temperature: $T = 23\,^\circ C$
- Altitude = 800 m
- DO concentration to be maintained in the liquid: $C_L = 1.5$ mg/L

Example 5.3 (*Continued*)

Solution:

Adopt the following values for the parameters of Equation 5.25:

C_s (20 °C) = 9.2 mg/L, column 0 m altitude, for T = 20 °C
C_s = 8.7 mg/L, column 0 m altitude, for T = 23 °C
α = 0.90 (see comments for Equation 5.24)
β = 0.95 (see comments for Equation 5.23)
θ = 1.024 (see comments for Equation 5.21)
According to Equation 5.8 the value of f_H is:

$$f_H = 1 - \frac{\text{altitude}}{9450} = 1 - \frac{800}{9450} = 0.92$$

According to Equation 5.25 the value of OTR standard is:

$$OTR_{standard} = \frac{OTR_{field}}{\dfrac{\beta.f_H.C_s - C_L}{C_s(20\,°C)}.\alpha.\theta^{T-20}}$$

$$= \frac{100}{\dfrac{0.95 \times 0.92 \times .8.7 - 1.5}{9.2}.0.9 \times 1.024^{23-20}} = \frac{100}{0.62}$$

$$= 161 kgO_2/h$$

The final results are:

OTR_{field} = 100 kgO$_2$/h (given in the problem)
$OTR_{standard}$ = 161 kgO$_2$/h
Ratio $OTR_{field}/OTR_{standard}$ = 100/161 = 0.62 = 62%

Therefore, it can be seen that in the field the aeration system is capable of supplying only 62% of the capacity under standard conditions. For this reason, to obtain the value of 100 kgO$_2$/h in the field, a system that supplies 161 kgO$_2$/h under standard conditions must be specified.

5.6 OTHER AERATION COEFFICIENTS

5.6.1 Oxygenation efficiency

The *oxygenation efficiency* (OE) represents the oxygen transfer rate (kgO$_2$/h) per unit power consumed (kW), and is expressed in the units of kgO$_2$/kWh.

$$OE = \frac{OTR_{standard}}{P} \tag{5.27}$$

where:
$\quad$ OE = oxygenation efficiency (kgO$_2$/kWh)
$\quad\;$ P = power consumed (kW)

Tables 5.5 and 5.6 (Sections 5.7 and 5.8) present typical values of the oxygenation efficiency (standard conditions) for the most commonly used aeration systems.

The power consumed is related to the voltage and amperage by (Boon, 1980):

$$P = \frac{\sqrt{3}.\text{volt}.i.\cos\phi}{1000} \tag{5.28}$$

where:
$\quad$ volt = voltage (V)
$\qquad$ i = current intensity (A)
$\quad$ cosϕ = power factor (−)

In the case of mechanical aeration, it must be made clear if the power consumed does or does not include the efficiencies of the motor and the reducer.

In the case of diffused air aeration, the power required by the blowers can be expressed in terms of airflow and the pressure to be overcome by (Pöpel, 1979):

$$P = \frac{Q_g.\rho.g.(d_i + \Delta H)}{\eta} \tag{5.29}$$

where:
$\quad$ P = required power (W)
$\quad$ ρ = density of the liquid (1000 kg/m^3)
$\quad$ g = acceleration due to gravity (9.81 m/s^2)
$\quad$ d_i = immersion depth of the diffusers (m)
$\quad$ ΔH = head loss in the air distribution system (m)
$\quad$ η = efficiency of the motor and blower (−)

5.6.2 Oxygen transfer efficiency

In diffused air aeration systems, the ratio of oxygen utilisation (ROU) is expressed as the quantity of oxygen absorbed per m^3 of air applied (Pöpel, 1979):

$$\boxed{\text{ROU} = \frac{\text{OTR}_{\text{standard}}}{Q_g}} \tag{5.30}$$

where:
$\quad$ ROU = ratio of oxygen utilisation (kgO$_2$ absorbed/m^3 air applied)
$\quad$ Q_g = air flow (m^3/h)

The *standard oxygen transfer efficiency* (**SOTE**) represents the oxygen absorption efficiency in percentage terms. Since dry air contains 20.95% of oxygen on a volumetric basis, the molecular weight of oxygen is 32 g/mol and a gas occupies 0.0224 m³/mol, then the concentration of oxygen in the gas is $= 0.2095 \times 32/0.0224 = 299$ gO$_2$/m³. The SOTE can therefore be calculated through (Pöpel, 1979):

$$SOTE = 100.\frac{ROU}{C_g} = 100.\frac{ROU(g/m^3)}{299(g/m^3)} = 0.334.ROU\ (\%) \qquad (5.31)$$

where:

SOTE = standard oxygen transfer efficiency (%)

In order to take into consideration the depth of the diffuser d_i, the ratio of oxygen utilisation can be related to d_i, leading to the parameter *ratio of oxygen utilisation per unit immersion* (ROU/d_i), expressed as gO$_2$/m³.m.

Table 5.5 (Section 5.8) presents typical SOTE values (standard conditions) for the more commonly used diffused air aeration systems.

5.6.3 Power level

The basic functions of an aeration system in most of the aerated wastewater treatment systems are:

- oxygenation of the wastewater under treatment
- liquid mixing, in order to maintain the biomass in suspension

To achieve the second objective, it is necessary to introduce a power per unit volume sufficient to avoid the settlement of the solids. This relation is represented through the concept of the *power level* (PL or φ), expressed as:

$$PL = \frac{P}{V} \qquad (5.32)$$

where:

PL = power level (W/m³)
 P = power input (W)
 V = reactor volume (m³)

The greater the power level, the greater the quantity of suspended solids that can remain dispersed in the liquid medium (see Table 5.4). The values presented in the table are only estimates, since the mixing intensity also depends on the number and distribution of aerators (in the case of mechanical aeration) and on the size and geometry of the tank.

Owing to the higher suspended solids concentrations in activated sludge reactors, the power level to be adopted should usually be higher than 10 W/m³.

Table 5.4. Suspended solids concentrations
that can be maintained dispersed in the liquid
as a function of the power level

Power level (W/m^3)	SS (mg/L)
0.75	50
1.75	175
2.75	300

Source: Eckenfelder (1980)

Example 5.4 illustrates the calculation of the various aeration coefficients for a diffused air system. The calculation for a mechanical aeration system is simpler because the main coefficients are only OE and PL.

Example 5.4

Determine the main parameters of the following diffused air aeration system (medium size bubbles)

- Net reactor volume : $V = 500 \ m^3$
- Air flow: $Q_g = 0.6 \ m^3/s$
- Immersion depth of the diffusers: $d_i = 4.0 \ m$ (tank height)
- Head loss in the air distribution system: $\Delta H = 0.4 \ m$
- Standard oxygen transfer rate: $OTR_{standard} = 60 \ kgO_2/h$
- Efficiency of the motor and blower: $\eta = 0.60$

Solution:

a)   Ratio of oxygen utilisation

$$Q_g = 0.6 \ m^3/s \times 3{,}600 \ s/h = 2{,}160 \ m^3/h$$

From Equation 5.30:

$$ROU = \frac{OTR_{standard}}{Q_g} = \frac{60{,}000 \ gO_2/h}{2{,}160 \ m^3/h} = 27.8 \ gO_2/m^3 air$$

b)   Ratio of oxygen utilisation per unit immersion

$$\frac{ROU}{d_i} = \frac{27.8 \ gO_2/m^3 air}{4.0 \ m} = 7.0 \ gO_2/m^3.m$$

c)   Standard oxygen transfer efficiency

From Equation 5.31:

$$SOTE = 0.334.ROU = 0.334 \times 27.8 = 9.3\%$$

Example 5.4 (Continued)

d)   Required power

From Equation 5.29:

$$P = \frac{Q_g \cdot \rho \cdot g \cdot (d_i + \Delta H)}{\eta} = \frac{0.6 \times 1000 \times 9.81 \times (4.0 + 0.4)}{0.60}$$

$$= 43,164 \text{ W} = 43.2 \text{ kW}$$

e)   Oxygenation efficiency

From Equation 5.27:

$$OE = \frac{OTR_{standard}}{P} = \frac{60 \text{kgO}_2/\text{h}}{43.2 \text{ kW}} = 1.39 \text{ kgO}_2/\text{kWh}$$

f)   Power level

From Equation 5.32:

$$PL = \frac{P}{V} = \frac{43,164 \text{ W}}{500 \text{ m}^3} = 86 \text{ W/m}^3$$

5.7 MECHANICAL AERATION SYSTEMS

The main mechanisms of oxygen transfer by mechanical surface aerators are (Malina, 1992):

- Atmospheric oxygen transfer to the droplets and the fine films of liquid sprayed in the air
- Oxygen transfer at the air-liquid interface, where the falling drops enter into contact with the liquid in the reactor
- Oxygen transfer by air bubbles transported from the surface to the bulk of the liquid medium

The more commonly used mechanical aerators can be grouped according to:

Classification as a function of the rotation shaft:

- *vertical shaft aerators*
  - low speed, radial flow
  - high speed, axial flow
- *horizontal shaft aerators*
  - low speed

MECHANICAL AERATORS

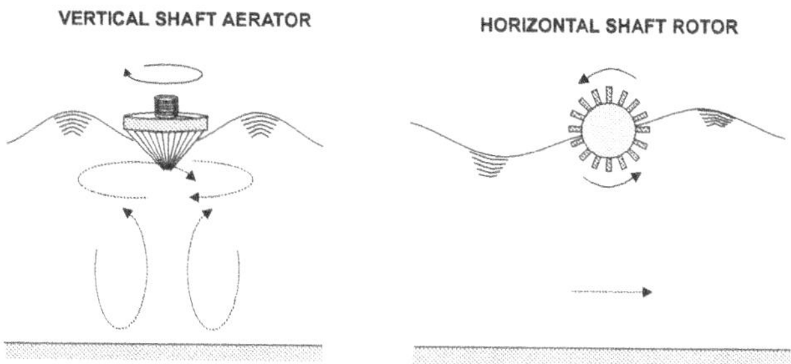

Figure 5.5.   Schematic representation of vertical and horizontal shaft aerators

Classification as a function of the supporting:

- *fixed aerators*
- *floating aerators*

Figure 5.5 shows schematically mechanical aerators with vertical and horizontal shafts. Table 5.5 presents a comparison between the three main types: (a) vertical shaft – low speed, (b) vertical shaft – high speed and (c) horizontal shaft.

The power of mechanical aerators usually varies between 5 HP and 100 HP, although, in special conditions, lower and higher values can be found.

In mechanical aerators, the submergence of the impellers in relation to the water level is a very important aspect in terms of oxygen transfer and energy consumption. The following situations can occur:

- *Adequate submergence.* The performance is optimal. There is good turbulence and absorption of air with relation to the oxygen consumption.
- *Submergence above the optimal.* The unit tends to function more as a mixer than as an aerator. The energy consumption increases without being accompanied by a substantial increase in the oxygen transfer rate.
- *Submergence below the optimal.* Only a surface spray is formed in the vicinity of the aerator, without creating an effective turbulence. The energy consumption and the oxygen transfer rate decrease.

The installation of the aerator must follow the manufacturer's instructions. Besides this, local tests should be carried out in order to obtain the optimal submergence in the reactor in question.

In many activated sludge plants, the oxygen transfer rate can be varied in such a way as to adjust itself to the variations in the oxygen utilisation rate. The variation can be manual or automated, by means of timers or sensors for dissolved oxygen in the reactor. Listed below are some of the most common forms of varying the

Table 5.5. Characteristics of the main mechanical aeration systems

Type of aerator	Characteristics	Application	Components	Advantages	Disadvantages	Standard oxygenation efficiency (kgO_2/kWh)
Low speed, radial flow	Similar to a high flow and low head pump. The flow of the liquid in the tank is radial in relation to the axis of the motor. Most of the oxygen absorption results from an induced hydraulic jump. Rotation speed 20–60 rpm.	Activated sludge and variants. Aerobic digesters. Large aeration units with depths up to 5 m.	Motor, reducer, impeller. Fixation units (bridges or platforms) for the fixed aerators (more common).	High oxygen transfer. Good mixing capacity. Flexibility in the design of the tank. High pumping capacity. Easy access for maintenance.	High initial costs. Careful maintenance of the reducers is necessary.	1.4 – 2.0
High speed, axial flow	Similar to a high flow and low head pump. The flow of the liquid pumped is upwards and follows the axis of the motor, passing through the volute before reaching a diffuser, where it is dispersed perpendicularly to the axis of the motor in the form of a spray. Most of the oxygen absorption occurs due to spray and turbulence. Rotation speed: 900 – 1400 rpm.	Activated sludge and variants. Aerobic digesters. Aerated lagoons.	Motor, impeller, float (a reducer is not needed).	Lower initial costs. Easily adjustable to variations in the water level. Flexible operation.	Difficult access for maintenance. Lower mixing capacity. Oxygen transfer not very high.	1.0 – 1.4
Horizontal shaft	The rotation is around the horizontal shaft. When rotor is rotating, a large number of fins perpendicular to the shaft cause aeration by spray and incorporation of air, besides providing the horizontal movement of the liquid in the reactor. Rotation speed: 20 – 60 rpm.	Activated sludge oxidation ditches (depth less than 2.5 m)	Motor, reducer, rotor.	Moderate initial cost. Easy to fabricate locally. Easy access for maintenance.	Limited shape of the tank. Low depth requirement. Possible problems with long shaft rotors. Oxygen transfer not very high.	1.2 – 2.0

Source: Arceivala (1981), Qasim (1985), Metcalf & Eddy (2002), Malina (1992), WEF & ASCE (1992)

oxygen transfer rate in mechanical aerator systems:

- switch on and off certain aerators
- vary the rotation speed of the aerators
- vary the submergence of fixed aerators through the alteration of the level of the outlet weir (change in the water level)
- vary the submergence through the change of the level of the aerator shaft

5.8 DIFFUSED AIR AERATION SYSTEMS

The diffused air aeration system is composed of diffusers submerged in the liquid, air distribution piping, air transport piping, blowers, and other units through which the air passes. The air is introduced close to the bottom of the tank and the oxygen is transferred to the liquid medium while the bubble rises to the surface.

The main diffused air systems can be classified according to the porosity of the diffuser and the size of the bubble produced:

- **porous diffuser** (*fine* and *medium bubbles*): plate, disc, dome, tube (ceramic, plastic, flexible membrane)
- **non-porous diffuser** (*coarse bubbles*): nozzles or orifices
- **other systems**: jet aerator, aspirating aerator, U-tube aerator

Figure 5.6 presents an schematics of the aeration by porous diffusers and aspirating devices. Aspirating devices have an impeller at the lower end (immersed in the liquid), which, when rotating, create a negative pressure, sucking in atmospheric air through a slot situated at the upper end (outside the liquid). Air is diffused into the liquid medium in the form of small bubbles, which are responsible for the oxygenation and mixing of the liquid mass. The aspirating aerators are presented in some texts as mechanical aerators, since they have motors that rotate outside the liquid, and in other texts as diffused air aerators, because they generate air bubbles in the liquid medium.

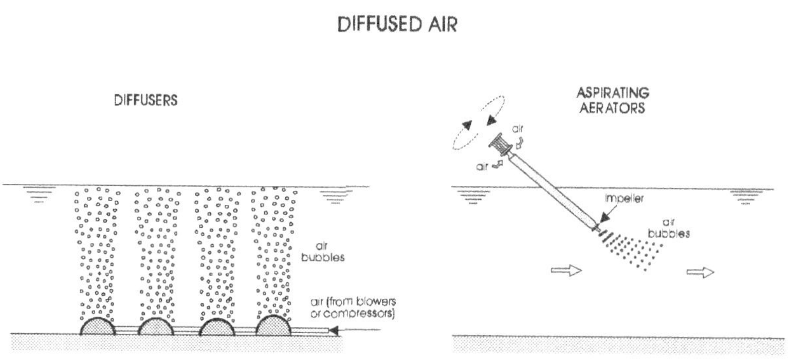

Figure 5.6. Diffused air aeration by porous diffusers and by aspirating devices

The diameters of the bubbles considered in the classification of the aeration type are (ABNT, 1989):

- *fine bubble*: diameter less than 3 mm
- *medium bubble*: diameter between 3 and 6 mm
- *coarse bubble*: diameter greater than 6 mm

In general, the smaller the size of the air bubbles, the greater the surface area available for gas transfer, that is, the greater the oxygenation efficiency. For this reason, aeration systems with fine bubbles are the most efficient in the transfer of oxygen.

The oxygen transfer efficiency of the porous diffusers decreases with the use, due to the internal or external clogging. The internal clogging is due to impurities in the air that are not removed by the filter. The external clogging is due to bacterial growth on the surface, or the precipitation of inorganic compounds.

The oxygen transfer rate can be changed to adjust itself to the oxygen consumption through the control of the blowers and the air distribution system, thus allowing energy savings.

Table 5.6 presents the characteristics of the main diffused air aeration systems.

5.9 AERATION TESTS

Wastewater treatment plants are designed based on a desired oxygen transfer rate from the aeration system. Normally this transfer rate, whether expressed in standard conditions or in field operating conditions, is part of the specification for purchasing the aeration equipment. Unfortunately, it has not been a common practice the undertaking of aeration tests to verify if the equipment being supplied satisfies the required oxygen demand. Even with the tests carried out in the manufacturer's laboratory, the transformation of the standard condition values to the real situation in the treatment plant is difficult, because of the various influencing factors, such as the tank shape, number and placing of the aerators and others.

In the existing treatment plants it is very important to know the oxygenation capacity of the installed equipment. In the same way that the influent quality is monitored in order to allow the estimation of oxygen **consumption** (BOD), it is equally important to have the knowledge of *the real capacity of oxygen **production** available in the reactor under operational conditions*.

This aspect becomes even more important, considering that there is an optimal operating point that leads to the greatest oxygen transfer efficiency (mass of O_2 supplied per unit of energy consumed). For instance, in reactors with mechanical aeration, this point is obtained at a certain submergence of the aerators, which can be achieved through the adjustment of the level of the outlet weir (which may be also variable during the day). Hence, it is important that aeration tests be carried out under operational conditions, aiming at determining the level of the outlet weir that leads to the supply of the *required O_2 mass* within the *greatest possible transfer efficiency*. Considering that the greatest energy costs in a treatment plant are related with aeration, the economy resulting from this procedure can be considerable.

Table 5.6. Characteristics of the main diffused air systems

Aeration type	Characteristics	Application	Advantages	Disadvantages	Average standard oxygen transfer efficiency (%)	Standard oxygenation efficiency (kgO_2/kWh)
Fine bubbles	The bubbles are produced in plates, discs, tubes or domes, made of a ceramic, glass or resin medium	Activated sludge	High oxygen transfer. Good mixing capacity. High operational flexibility through the variation of the airflow.	High initial and maintenance costs. Possibility of clogging of the diffusers. Air filters are necessary.	10–30	1.2–2.0
Medium bubbles	The bubbles are produced in perforated membranes or perforated tubes (coated stainless steel or plastic)	Activated sludge	Good mixing capacity. Reduced maintenance costs.	High initial costs. Air filters could be necessary.	6–15	1.0–1.6
Coarse bubbles	The bubbles are produced in orifices, nozzles, or injectors.	Activated sludge	No clogging. Low maintenance costs. Competitive initial costs. Air filters are not necessary.	Low oxygen transfer. High-energy requirements.	4–8	0.6–1.2
Aspirating aerators	The bubbles are produced by a propeller rotating at high speed at the bottom of a tube, which sucks in atmospheric air through the orifice at the upper end of the tube.	Aerated lagoons, activated sludge	No clogging. Air filters are not necessary. Conceptual simplicity. Maintenance relatively simple.	Lower oxygenation efficiency compared to mechanical aeration or fine bubble systems.	–	0.6–1.2

Source: Qasim (1985), Metcalf & Eddy (1991), Malina (1992), WEF & ASCE (1992)

The aeration tests in an activated sludge plant can be done according to one of the following methods:

- *test with clean water*
  - steady-state method
  - unsteady-state method
- *test under operating conditions*
  - steady-state method
  - unsteady-state method

The **clean water test** requires that the tank is emptied and filled with clean water. The dissolved oxygen (DO) in the medium is removed through the addition of sodium sulphite in the reactor, with the aerators switched off. The aeration capacity is calculated based on the rate of increase in the DO concentration after turning off the aerators. This method is expensive due to the requirement of large volumes of treated water and chemical products, besides being often inpractical in operating plants.

The **test under operating conditions** is accomplished with the reactor in operating conditions containing the mixed liquor. The oxygen consumption results from the respiration of the biomass in the mixed liquor. This method is cheaper, since it does not require the addition of external products, and can be done with the treatment plant in operation. Even though the results can be less accurate than those with clean water (if the DO concentration remains low, even with the aerators turned on), the values obtained provide a direct indication of the Oxygen Transfer Rate really available in the system.

The **steady-state method** is that in which all the conditions in the reactor are constant (or practically constant), that is, there are no variations during the test period. Under these conditions, the oxygen consumption is equal to its production. The aeration capacity can be then estimated through the determination of the oxygen utilisation rate by the biomass.

The **unsteady-state** method (reaeration method) consists in turning on the aerators, aiming at increasing the dissolved oxygen concentration in the medium. The Oxygen Transfer Rate is associated with the measured rate of increase in the DO concentration.

The references Boon (1980), de Korte and Smits (1985), Stephenson (1985), WPCF & ASCE (1988), ASCE (1990), WEF & ASCE (1992) and von Sperling (1993) present descriptions of the aeration tests. Under operating conditions (more frequent and practical situation), the tests can be carried out in the following simplified way:

Steady-state method:

1. determination of the oxygen consumption rate (r)
2. production rate $=$ consumption rate
3. calculation of the oxygen transfer coefficient $K_L a$:

$$K_L a = \frac{r}{(C_s - C)} \qquad (5.33)$$

Unsteady-state method:

1. determination of the oxygen consumption rate (r)
2. production rate: determination by the DO trajectory after switching on the aerators
3. calculation of the oxygen transfer coefficient K_La:

$$\boxed{C = C_\infty - (C_\infty - C_o).e^{-K_La.(t-t_o)}}$$ (5.34)

$$\boxed{C_\infty = C_s - (r/K_La)}$$ (5.35)

The hydraulic regime has a large influence on the determination of the coefficient K_La. Table 5.7 presents a summary of the formulas to be used in the tests under operating conditions. In some of them K_La is explicit, while in others K_La should be obtained by regression analysis with the various values of the pairs DO × t (von Sperling, 1993). Even though regression analysis with the original equation 5.34 is frequently the preferred method to estimate the coefficient K_La, there are some alternative approaches based on the transformation of the basic equation. Example 5.5 presents a process based on the adoption of logarithms on both sides of Equation 5.34.

Example 5.5

Determine the K_La coefficient of the aeration system of a reactor with clean water, which has had the dissolved oxygen previously removed by the addition of sodium sulphite. After the removal of the DO, there was no further oxygen demand. After turning on the aerator, the DO values as a function of time were measured, and are presented in the table below. The saturation concentration of DO in the liquid, as a function of temperature and altitude was estimated as 8.4 mg/L.

t (s)	0	120	240	360	480	600	720	840
DO (mg/L)	0	1.0	1.8	2.5	3.2	3.8	4.3	4.8

Solution:

Applying the logarithm on both sides of Equation 5.34, after rearrangement:

$$\ln\left(\frac{C_s - C}{C_s - C_o}\right) = -K_La.(t - t_o)$$

In a scatter plot on the Y-axis of the various values of $\ln[(C_s - C)/(C_s - C_o)]$ and on the X-axis the values of $(t - t_o)$, K_La corresponds to the slope of the line

---

Example 5.5 (Continued)

of best fit. The data necessary for the construction of the graph are calculated below:

t (s)	0	120	240	360	480	600	720	840
$\ln[(C_s - C)/(C_s - C_o)]$	0.00	−0.13	−0.24	−0.35	−0.48	−0.60	−0.72	−0.85

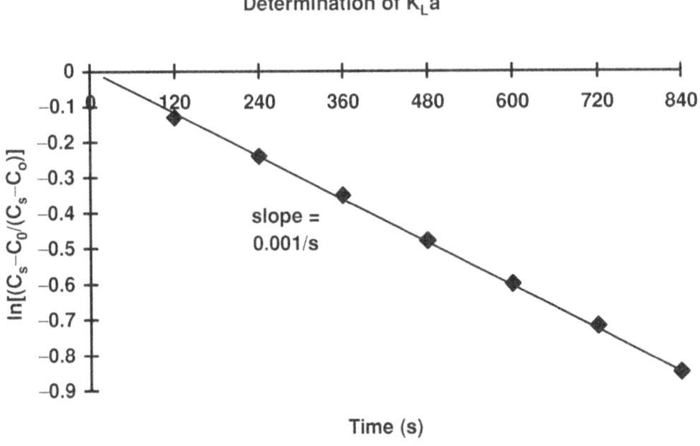

The slope of the line of best fit is 0.001 s^{-1}. This is the value of K_La determined in the experiment.

5.10 GRAVITY AERATION

In some cases, it may be interesting to increase the DO concentration in the effluent from a wastewater treatment plant, aiming at reaching higher concentrations in the water body, at the effluent-river mixing point. A simple way to achieve this is through the process of **gravity aeration**, used in some treatment plants.

Gravity aeration takes place in weirs or in steps in aeration cascades. Effluents from several wastewater treatment processes may benefit from an increased DO concentration. *However, it should be kept in mind that aeration is a gas transfer process: anaerobic effluents submitted to gravity aeration tend to release H_2S, which may cause bad odour and corrosion problems.*

The following text, based on von Sperling (1983b) and Pöpel (1979), describes the principles and application of gravity aeration.

The principle of gravity aeration is the use of the potential energy of the water to create gas–liquid interfaces for an efficient gas transfer. When the water passes over the crest of the weir or cascade, two different gas transfer mechanisms may occur:

- exposure of the water to the surrounding air
- exposure of the air to the water mass

Table 5.7. Formula for the determination of $K_L a$ according to various hydraulic regimes

Hydraulic regime	Method	Volume of the reactor	Formula
Complete-mix reactor	Steady state	Total	$K_L a = \dfrac{r}{(C_s - C)}$
	Unsteady state	Total	$C = C_\infty - (C_\infty - C_0).e^{-K_L a.(t - t_0)}$ (nonlinear regression)
			$C_\infty = C_s - (r/K_L a)$
Plug-flow reactor	Steady state	Complete mixing zone	$K_L a = \dfrac{r}{(C_s - C)}$
	Unsteady state	Complete mixing zone	$C = C_\infty - (C_\infty - C_0).e^{-K_L a.(t - t_0)}$ (nonlinear regression)
			$C_\infty = C_s - (r/K_L a)$
Carrousel	Steady state (DO increase)	Complete mixing zone	$K_L a = \dfrac{Q.(C - C_i) + r.V}{V.(C_s - C)}$
	Steady state (DO consumption)	Total (1 aerator turned on)	$K_L a = \dfrac{r}{(C_s - C)}$
	Unsteady state	Total	$C = C_\infty - (C_\infty - C_0).e^{-K_L a.(t - t_0)}$ (nonlinear regression)
			$C_\infty = C_s - (r/K_l a)$
Pasveer ditch	Steady state (DO consumption)	Total (1 aerator turned on)	$K_L a = \dfrac{r}{(C_s - C)}$
	Unsteady state	Total	$C = C_\infty - (C_\infty - C_0).e^{-K_L a.(t - t_0)}$ (nonlinear regression)
			$C_\infty = C_s - (r/K_L a)$

The first mechanism relates to the *exposure of the water to the surrounding air*, which occurs during the free fall. If the fall height H is known, the average air exposure time $[t = (2H/g)^{0.5}]$ can be estimated, which allows an evaluation of the gas transfer coefficients. The configuration of the water fall crest influences the aeration, because the subdivision of the flow into several jets increases the air-water contact area, enabling an increased efficiency of the gas transfer operation.

The second mechanism refers to the *exposure of air to the water mass*, exactly the reverse of the first phenomenon. It occurs due to the submergence of the flow into the bulk of the liquid located on the base of the waterfall, causing significant amounts of air to be absorbed. The incorporated air is then dispersed under the form of bubbles in the liquid, leading to an intense gas transfer. The amount of air absorbed in the second mechanism depends primarily on the velocity [$v = (2gH)^{0.5}$] of the jet passing through the surface of the downstream water. Consequently, the gas transfer is substantially determined by the height of the fall, in a much more significant manner than in the first mechanism mentioned.

Besides that, the depth of the receiving water influences the amount of gas transferred: the deeper the jet can submerge into the water mass, the larger the specific surface area and the longer the contact time between the bubbles and the water. For an optimal utilisation of this effect, the depth should be such that the final velocity of the jets prior to reaching the bottom is equal to the upward velocity of the bubbles produced.

In general terms, it is understood that the first mechanism is efficient for gas release, and the second for gas absorption. Thus, for example, hydrogen sulphide has better release conditions during the free fall phase, while oxygen is mostly absorbed after the submergence of the flow into the downstream water. In summary:

- *water surrounded by air*: predominance of gas release
- *air surrounded by water*: predominance of gas absorption

A large part of the oxygen absorption is also caused by the shock of the water jets against obstacles, allowing the subdivision of the falling liquid mass, thus increasing the exposure area. In addition to that, if the water does not fall freely, but attached to the face of the waterfall or steps, the aeration will be significantly reduced.

The effluent (downstream) oxygen concentration can be estimated based on the coefficient of gas transfer, named **efficiency coefficient (K)** in the case of gravity aeration. Knowing the K value for the water fall at issue, the effluent DO concentration can be estimated for different conditions of saturation and influent concentrations:

$$\boxed{C_e = C_o + K.(C_s - C_o)} \qquad (5.36)$$

where:

C_o = influent (upstream) oxygen concentration (mg/L)
C_e = effluent (downstream) oxygen concentration (mg/L)
C_s = oxygen saturation concentration (mg/L)
K = efficiency coefficient (dimensionless)

The K coefficient is specific and constant for each aeration system (in this case, each water fall), provided that certain conditions, such as the influent flow, remain constant. In an existing waterfall, the K coefficient can be obtained by rearranging Equation 5.36.

$$K = \frac{C_e - C_o}{C_s - C_o} \qquad (5.37)$$

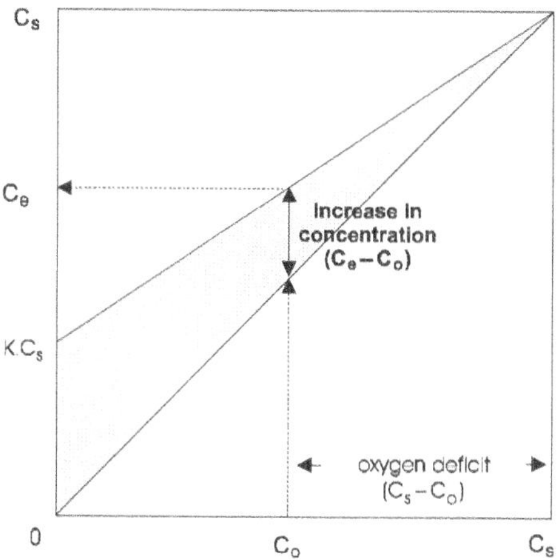

Figure 5.7.   Estimated DO gains in gravity aeration

The conceptual graphic representation of Equation 5.36 is shown in Figure 5.7. In this figure, the C_e value is obtained from the K coefficient and the C_o and C_s concentrations. The figure emphasises the large influence of the oxygen deficit $(C_s - C_o)$ on the increase of the oxygen concentration. For the case in which $C_o = 0$, the gain in the oxygen concentration will be $C_e - C_o = K.C_s$. Therefore, the K coefficient establishes the fraction of the saturation concentration to be gained in aerating a water completely devoid from oxygen. K is always lower than 1. It is also observed that, the higher (closer to C_s) the influent concentration C_o, the lower the oxygen deficit and, consequently, the lower the increase in the oxygen concentration $(C_e - C_o)$. This can be the case of effluents from facultative and maturation ponds containing high DO contents.

The literature presents some empirical formulas for the determination of K according to the fall height (each single step) and other conditions (Table 5.8).

Equation 5.36 applies to *each step* or level of the cascade. For a system consisting of several free falls (e.g. steps), the overall K of the sequence of steps may be determined based on the individual K for each step according to:

$$K = 1 - [(1 - K_1).(1 - K_2) \cdots (1 - K_n)] \qquad (5.38)$$

where:

K_1 = efficiency coefficient of the first free fall
K_2 = efficiency coefficient of the second free fall
K_n = efficiency coefficient of the nth free fall
K = overall efficiency coefficient of the system

Table 5.8. Formulas to determine the efficiency coefficient K for gravity aeration in weirs and cascades

Author	K coefficient	Coefficients of the equation
Barret, Gameson, and Ogden (apud Pöpel, 1979)	$K = P.(1 + 0.046.T).H$	$P = 0.45$ (clean water) $P = 0.36$ (polluted water) $P = 0.29$ (sewage)
Kroon and Schram (apud von Sperling, 1983b)	$K = R.H$	$R = 0.40$ $R = 0.64$ (in case of weirs with more than 4 jets per linear metre and falls lower than 0.70 m)
Parkhurst and Pomeroy (1972)	$K = 1 - e^{-F.H}$	$F = 0.53$ m^{-1} (clean water over weirs) $F = 0.41$ m^{-1} (slightly polluted water over weirs) $F = 0.28$ m^{-1} (treatment effluents over weirs)

Source: von Sperling (1983b)
H = height of each free fall (m)
T = temperature of the liquid (°C)

In the frequent case in which the steps have the same height, Equation 5.38 is simplified to:

$$K = 1 - (1 - K_1)^n \qquad (5.39)$$

where:
n = number of equal free falls in the aeration system

Example 5.6

Estimate the effluent concentration of a step aeration, based on the following data:

- Influent DO concentration to the sequence of steps (effluent from the wastewater treatment plant): $C_o = 3.0$ mg/L
- DO saturation concentration: $C_s = 8.5$ mg/L
- $T = 20\,°C$
- Height of each step: $H = 0.30$ m
- Number of steps: $n = 5$

Solution:

a)   Determination of the K coefficient for each step

Based on the formulas on Table 5.8:

- Barret, Gameson, and Ogden

With $P = 0.33$ (adopted), $T = 20\,°C$ and $H = 0.30$ m:
$K = P.(1 + 0.046.T).H = 0.33 \times (1 + 0.046 \times 20) \times 0.30 = 0.19$

Example 5.6 (Continued)

- Kroon and Schram

With R = 0.40 (adopted) and H = 0.30 m:

K = 0.40 × 0.30 = 0.12

- Pomeroy

With F = 0.32 (adopted) and H = 0.30 m:

$K = 1 - e^{-F.H} = 1 - e^{-0.32 \times 0.30} = 0.09$

Adopt an intermediate value, such as: K = 0.13.

b)   Determination of the overall K coefficient of the step aeration system

Since the steps are equal, Equation 5.39 is used, with n = 5 steps:

$$K = 1 - (1 - K_1)^n = 1 - (1 - 0.13)^5 = 0.50$$

c)   Determination of the effluent DO concentration

From Equation (5.36):

$$C_e = C_o + K.(C_s - C_o) = 3.0 + 0.50 \times (8.5 - 3.0) = 5.8 \text{ mg/L}.$$

Therefore, the DO has been increased by 2.8 mg/L, and the concentration raised from 3.0 mg/L to 5.8 mg/L.

d)   Calculation of the DO gain should the concentration of influent DO be 1.0 mg/L and 5.0 mg/L

For C_o = 1.0 mg/L:
$C_e = C_o + K.(C_s - C_o) = 1.0 + 0.50 \times (8.5 - 1.0) = 4.8$ mg/L (gain of 3.8 mg/L)

For C_o = 5.0 mg/L:
$C_e = C_o + K.(C_s - C_o) = 5.0 + 0.50 \times (8.5 - 5.0) = 6.8$ mg/L (gain of 1.8 mg/L)

The influence of the influent concentration, that is, of the DO deficit, is evident from these calculations.

e)   Comments

The gain of DO can be optimised by trying different combinations of numbers of steps and individual heights of each step, within the total height available to allocate the step aeration system.

References

ABNT (1989). *Projeto de estações de tratamento de esgotos.* NBR-570 (in Portuguese).

ARCEIVALA, S.J. (1981). *Wastewater treatment and disposal.* Marcel Dekker, New York. 892 p.

ASCE (1990). *ASCE Standard. Measurement of oxygen transfer in clean water.* ANSI/ASCE, 2-90. 66 p.

BENEFIELD, L.D., RANDALL, C.W. (1980). *Biological process design for wastewater treatment.* Prentice-Hall, EUA. 526 p.

BOON, A.G. (1980). Measurement of aerator performance. In: *Symposium on the profitable aeration of wastewater.* London, 25/04/1980.

BRANCO, S.M. (1976). Biologia da poluição. In: CETESB (1976*). Ecologia aplicada e proteção do meio ambiente.* São Paulo (in Portuguese).

BRANCO, S.M. (1978) *Hidrobiologia aplicada à engenharia sanitária.* São Paulo, CETESB. 620 p (in Portuguese).

CATUNDA, P.F.C., VAN HAANDEL, A.C. (1987). Activated sludge settlers: design and optimization. *Water Science and Technology,* **19**, pp. 613–623.

CHERNICHARO, C.A.L. (1997). Reatores anaeróbios. Departamento de Engenharia Sanitária e Ambiental, UFMG, 245 p (in Portuguese).

DE KORTE, K., SMITS, P. (1985). Steady state measurement of oxygenation capacity. *Water Sci. Technol.,* v. 17, p 303–311.

DICK, R.I. (1972). Gravity thickening of sewage sludges. *Water Pollution Control,* **71**, pp. 368–378.

ECKENFELDER Jr, W.W. (1980). *Principles of water quality management.* Boston, CBI. 717 p.

ECKENFELDER Jr., W.W. (1989). *Industrial water pollution control.* McGraw Hill International.

ECKENFELDER, W.W., GRAU, P. (1992*). Activated sludge process design and control. Theory and practice.* Technomic Publishing Co, Lancaster, EUA. 268 p.

EPA, Environmental Protection Agency, Cincinatti (1993). *Nitrogen control.* Technology Transfer. 311 p.

GRADY, C.P.L., LIM, H. (1980). *Biological wastewater treatment: theory and application.* Marcel Dekker, New York.

194 References

HANDLEY, J. (1974). Sedimentation: an introduction to solids flux theory. *Water Pollution Control*, **73**, pp. 230–240.

HANISCH, B. (1980). *Aspects of mechanical and biological treatment of municipal wastewater*. Delft, IHE.

HORAN, N.J. (1990). *Biological wastewater treatment systems. Theory and operation*. John Wiley & Sons, Chichester. 310 p.

HUISMAN, L. (1978). *Sedimentation and flotation. Mechanical filtration*. 2. ed. Delft, Delft University of Technology.

IAWPRC (1987). *Activated sludge model No. 1*. IAWPRC Scientific and Technical Reports No. 1.

IWA (2000). *Activated sludge models ASM1, ASM2, ASM2d and ASM3*. IWA Scientific and Technical Report No. 9. IWA Publishing.

IWAI, S., KITAO, T. (1994). *Wastewater treatment with microbial films*. Technomic Publishing Co, Lancaster, EUA. 184 p.

KEINATH, T.M. (1981). Solids inventory control in the activated sludge process. *Water Science and Technology*, **13**, pp 413–419.

KEINATH, T.M., RYCKMAN, M.D., DANA, C.B., HOFER, D.A. (1977). Activated sludge – unified system design and operation. *J. Environ. Eng. Div., ASCE*, **103** (EE5), pp. 829–849.

KÖNIG, A. (1990). Capítulo 3. Biologia das lagoas. Algas. In: MENDONÇA, S.R. (1990). *Lagoas de estabilização e aeradas mecanicamente: novos conceitos*. João Pessoa. 385 p (in Portuguese).

LA RIVIÉRE, J.W.M. (1977). Microbial ecology of liquid waste treatment. *Advances in microbial ecology*, **1**, p. 215–259.

LA RIVIÉRE, J.W.M. (1980). *Microbiology*. Delft, IHE (lecture notes).

LETTINGA, G. (1995). Introduction. In: *International course on anaerobic treatment*. Wageningen Agricultural University / IHE Delft. Wageningen, 17–28 Jul 1995.

LETTINGA, G., HULSHOF POL, L.W., ZEEMAN, G. (1996). *Biological wastewater treatment. Part I: Anaerobic wastewater treatment*. Lecture notes. Wageningen Agricultural University, jan 1996.

LUBBERDING, H.J. (1995). Applied anaerobic digestion. In: *International course on anaerobic treatment*. Wageningen Agricultural University / IHE Delft. Wageningen, 17–28 Jul 1995.

MALINA, J.F. Biological waste treatment. In: *Seminário de transferência de tecnologia. Tratamento de esgotos*. ABES/WEF. Rio de Janeiro, 17–20 Aug 1992. p. 153–315.

MARAIS ,G.v.R., EKAMA, G.A. (1976). The activated sludge process. Part I – Steady state behaviour. *Water S.A.*, **2** (4), Oct. 1976. p. 164–200.

METCALF & EDDY (1991). *Wastewater engineering: treatment, disposal and reuse*. Metcalf & Eddy, Inc. 3. ed, 1334 pp.

METCALF & EDDY (2003). *Wastewater engineering: treatment and reuse*. McGraw Hill, 4th ed. 1819 p.

O'CONNOR, D.J., DOBBINS, W.E. (1958). Mechanism of reaeration in natural streams. *Journal Sanitary Engineering Division, ASCE*, **123**. p. 641–666.

ORHON, D., ARTAN, N. (1994). *Modelling of activated sludge systems*. Technomic Publishing Co, Lancaster, EUA. 589 p.

PARKHURST, J.D., POMEROY, R.D. (1972). Oxygen absorption in streams. *Journal Sanitary Eng. Div., ASCE*, **98** (1), Feb. 1972.

PÖPEL, H.J. (1979). *Aeration and gas transfer*. 2. ed. Delft, Delft University of Technology. 169 p.

QASIM, S.R. (1985). *Wastewater treatment plants: planning, design and operation*. Holt, Rinehart and Winston, New York.

ROUXHET, P.G., MOZES, N. (1990). Physical chemistry of the interface between attached microorganisms and their support. *Water Research*, **22** (1/2), pp. 1–16.

SAWYER, C.N., Mc CARTY, P.L. (1978). *Chemistry for environmental engineering.* 3. ed. New York, Mc Graw-Hill, Inc. 532 p.

SILVA Jr., C., SASSON, S. *Biologia 2: Seres vivos, estrutura e função.* (Cesar e Sezar). Atual Editora, São Paulo, 2ª ed. 382 p (in Portuguese).

SILVA, S.A., MARA, D.D. (1979). *Tratamentos biológicos de águas residuárias: lagoas de estabilização.* ABES, Rio de Janeiro. 140 p (in Portuguese).

STEPHENSON, R. V. et al (1985). Performance of surface rotors in an oxidation ditch. *Journal of Environ. Eng. Div., ASCE,* v.111, n.1, p 79–91.

VAN HAANDEL, A.C., LETTINGA, G. (1994). *Tratamento anaeróbio de esgotos. Um manual para regiões de clima quente* (in Portuguese).

VIESMANN Jr, W., HAMMER, M.J. (1985). *Water supply and pollution control.* Harper and Row, Publ. New York. 4. ed. 797 p.

VON SPERLING, M. (1983a). A influência do tempo de retardo na determinação do co-eficiente de desoxigenação. *Engenharia Sanitária,* **22** (3), jul 1983. p. 375–379 (in Portuguese).

VON SPERLING, M. (1983b). *Autodepuração dos cursos d'água.* Dissertação de mestrado. DES-EE.UFMG, Belo Horizonte. 366 p (in Portuguese).

VON SPERLING, M. (1990). *Optimal management of the oxidation ditch process.* PhD Thesis, Imperial College, University of London, 1990. 371 p.

VON SPERLING, M. (1993). Determinação da capacidade de oxigenação de sistemas de lodos ativados em operação. In: *Congresso Brasileiro de Engenharia Sanitária e Ambiental,* **17**, Natal, 19–23 Setembro 1993, Vol. 2, Tomo I, pp. 152–167 (in Portuguese).

VON SPERLING, M. (1994). A new unified solids flux-based approach for the design of final clarifiers. Description and comparison with traditional criteria. *Water Science and Technology,* **30** (4). pp 57–66.

VON SPERLING, M. (1997*). Princípios do tratamento biológico de águas residuárias. Vol. 4. Lodos ativados.* Departamento de Engenharia Sanitária e Ambiental – UFMG. 428 p (in Portuguese).

VON SPERLING, M. (1999). Performance evaluation and mathematical modelling of co-liform die-off in tropical and subtropical waste stabilization ponds. *Water Research,* **33** (6). pp. 1435–1448.

VON SPERLING, M. (2002). Relationship between first-order decay coefficients in ponds, according to plug flow, CSTR and dispersed flow regimens. *Water Science and Tech-nology,* **45** (1). pp. 17–24.

VON SPERLING, M., FRÓES, C.M.V. (1999). Determination of the required surface area for activated sludge final clarifiers based on a unified database. *Water Research,***33** (8). pp. 1884–1894.

TCHOBANOGLOUS, G. & SCHROEDER, E.D. (1985*). Water quality: characteristics, modeling, modification.* Addison-Wesley, Reading, MA.

WANNER, J. (1994). *Activated sludge bulking and foaming control.* Technomic Publishing Company, 327 p.

WEF & ASCE (1992). *Design of municipal wastewater treatment plants.* Water Environ-ment Federation / American Society of Civil Engineers. 1592 pp.

WHITE, M.J.D. (1976). Design and control of secondary settlement tanks. *Water Pollution Control,* **75**, pp 459–467.

WILSON, F. (1981). *Design calculations in wastewater treatment.* E.&F.N.Spon, London. 221p.

WPCF & ASCE (1988). *Aeration. A wastewater treatment process.* 167 p.

IWA Publishing's authorised EU representative for General Product
Safety Regulations is Diane D'Arras, 15 rue Duret, 75116 Paris,
France, e-mail: safety@iwap.co.uk.

Printed and bound by CPI Group (UK) Ltd, Croydon, CR0 4YY
06/07/2026
02157563-0002